THERMAL ANALYSIS OF PRESSURIZED WATER REACTORS

Second Edition

L. S. Tong

Reactor Safety Research Division
U.S. Nuclear Regulatory Commission
Washington, D.C.

Joel Weisman

Department of Chemical and Nuclear Engineering
University of Cincinnati
Cincinnati, Ohio

Published by
AMERICAN NUCLEAR SOCIETY

Thermal analysis of pressurized water reactors

621.4834
T665

Tong, Long-sun, 1915-   Thermal analysis of
pressurized water reactors (second edition)
[C1979] by L. S. Tong and Joel Weisman. Pre-
pared under direction of the American Nuclear
Society. LaGrange Park, Ill., American Nuclear Society, 1979.

xvi, 440 p illus. 24 cm. (Monograph series)

1. Pressurized water reactors.   2. Heat—Transmission.
I. Weisman, Joel, 1928-   joint author.   II. American
Nuclear Society. III. United States. IV. Title.

TK9203.P7T6   621.48 ′34   79-54237
                                                    MARC

Library of Congress   71[71]

Library of Congress Catalog Card Number: 79-54237
International Standard Book Number: 0-89448-019-7

THERMAL ANALYSIS OF PRESSURIZED WATER REACTORS

To the memory of

Jay Weisman,

a young man full of promise.

Publisher's Foreword

This book is published as one of a continuing series in the American Nuclear Society's program for providing to the nuclear community and related fields authoritative information in monograph form. Authors and titles are selected to bring to print the most useful material in the more active areas of nuclear science and technology development. Advancing the peaceful uses of nuclear energy is a general objective of this publishing endeavor, while advancement of the professional interests of ANS members is a primary purpose.

The American Nuclear Society regards these publishing activities and their contributions to the growing achievements in nuclear energy applications as an essential obligation to the Society membership and the nuclear field.

Originally published in 1970, this second edition of *Thermal Analysis of Pressurized Water Reactors* by L. S. Tong and Joel Weisman has been completely revised and updated. References to the current technology and current literature have been included.

Norman H. Jacobson
Manager, ANS Publications

Preface to the Second Edition

The basic objective of this book is to present the principles underlying the thermal and hydraulic design of pressurized water reactors. In addition, the empirical data, engineering properties, and computer techniques required for design, but not available in conventional handbooks, are presented or referenced. This book is intended to provide an overview for nuclear engineering graduate students and to serve as a reference for engineers working in the nuclear power industry. This work is not intended to be a design manual.

While the foregoing objectives are essentially the same as those motivating preparation of the first edition, it was believed that a new edition was required since a number of significant advances have been made in the field since 1970. Improved understanding and correlations of fuel behavior, fluid dynamics, and heat transfer have been developed. Design techniques have become more sophisticated, and the design is more likely to be accomplished using a number of computer programs. In addition, the scope and depth of the safety analyses performed have been greatly extended.

Since the first edition of this monograph has apparently been found useful by a considerable number of readers, the basic format of the original monograph and as much of the text as is still relevant have been retained. Obsolete material has been removed and more recent correlations and design approaches have been added. Material dealing with computer-based design techniques and safety analysis has been considerably expanded. It is hoped that these revisions have enhanced the usefulness of this monograph.

L. S. Tong
Joel Weisman

January 1979

Contents

Chapter 1   Power Generation

Chapter 2   Fuel Elements

Chapter 3   Hydrodynamics

Chapter 4 Heat Transfer and Transport

Chapter 5   Thermal and Hydraulic Performance
of a Reactor Core

Nomenclature

This book follows the procedure of defining pertinent symbols after each equation. However, there are a number of symbols so frequently used that they have not been defined at each location of use. It is these symbols that are defined in this list.

a_l, a_v, a_{TP} = liquid, vapor, and two-phase sonic velocities, respectively

a_1 = sonic velocity, length/time

A = flow area, length^2

b = cladding radius, length

c_P, c_V = specific heats at constant pressure and volume, respectively, energy/mass deg

D = diameter, length

D' = extrapolated diameter of reactor, length

D_l = diffusion coefficient, length^2

D_e = equivalent diameter = $4A$/perimeter, length

f = flux depression factor, dimensionless

f = friction factor, dimensionless

F_Q^N, F_R^N, F_Z^N = nuclear hot-channel factors for overall heat flux, radial heat flux variation, and axial variation, respectively

$F_q, F_{\Delta H}$ = overall hot-channel factors for heat flux and enthalpy rise, respectively

$F_q^E, F_{\Delta H}^E$ = engineering hot-channel factors for heat flux and enthalpy rise, respectively

g_c = gravitational conversion factor, (mass/weight)(length/time^2)

g = gravitational acceleration, length/time^2

G = mass velocity, mass/time area

h = heat transfer coefficient, energy/time area deg

h_g = gap conductance, energy/time area deg

H = enthalpy, energy/mass

H_{fg} = enthalpy change on evaporation, energy/mass

H_V = enthalpy of saturated vapor, energy/mass

J = mechanical equivalent of heat, mechanical energy/thermal energy

k = thermal conductivity, energy/time length deg

k_{eff} = effective multiplication factor

k_{ex} = $k_{\text{eff}} - 1$

k_{∞} = infinite lattice multiplication factor

K = constant, form friction factor

L = length

L, L' = height and extrapolated height of reactor, respectively, length

p = probability, dimensionless

p = pitch, length

p' = perimeter, length

P = pressure, force/area

P' = power, energy/time

Pr = Prandtl Number = $c_P \mu / k$

q = total heat production rate, energy/time

q' = linear heat output, energy/time length

q'' = surface heat flux, energy/time length^2

q''_{crit} = critical heat flux, energy/time length^2

q''' = volumetric heat flux, energy/time length^3

Q = volumetric flow, volume/time

r = radius, length

Re = Reynolds Number = $De\ G/\mu$

S = slip ratio, vapor velocity/liquid velocity

t = time

t_1 = standard normal variate, dimensionless

T = temperature

u = radial or lateral velocity, length/time

u_l, u_v = superficial velocity of gas and vapor, respectively, length/time

U = internal energy, energy/mass

v = specific volume, volume/mass

v_c, v_f = cladding and fuel volumes, respectively, length^3

V = velocity, length/time

W = total flux, mass/time

w' = transverse-turbulent mixing flow, mass/length time

x, y, z = distance in coordinate directions, variables

X = quality = mass flow rate of vapor/total mass flow rate

Z = axial distance, length

α_a, α_b = thermal expansion coefficients of cladding and fuel, length/length deg

α = void fraction

α = parameter in escape probability calculation

γ = Poisson's ratio

γ = a constant

κ_0 = $(\Sigma_a/D_l)^{1/2}$

μ = attenuation coefficient, length^{-1}

μ = viscosity, mass/length time

$\bar{\mu}$ = mean value

ρ = density, mass/volume

σ = stress, force/area

σ = standard deviation

σ' = surface tension

Σ = macroscopic cross section, length^2

ϕ = neutron flux, neutrons/area time

1

POWER GENERATION

A pressurized water reactor (PWR) is a water-cooled nuclear reactor under sufficient pressure to prevent net steam generation at the core exit. In this chapter, we briefly delineate the major reactor systems falling under this definition and then describe, from a thermal analyst's viewpoint, the basic power distributions found within these reactors. The purpose of this discussion is to indicate the nature of the phenomena involved and some of the approximations that can be made.

1-1 REACTOR CONFIGURATIONS

1-1.1 Basic Concept

1-1.1.1 Overall System

In basic PWR design, coolant flowing through the core is cooled by heat exchange with a secondary fluid. The overpressure adequate to suppress bulk boiling is maintained by an electrically heated pressurizer. In a typical reactor system (Fig. 1-1) (Refs. 1 and 2), the primary coolant is circulated through the reactor core by one or more high pressure pumps. It proceeds to the heat exchanger where steam is generated and is then pumped back to the reactor core inlet. In the secondary system, feedwater is evaporated in the heat exchanger. Saturated steam goes through the turbine where it gives up its energy. It is then condensed in the condenser and returned to the heat exchanger by boiler feed pumps.

In commercial reactors built for central station power, the reactor core, contained in a large pressure vessel with a removable head, is moderated and cooled by light water. Shielding around the vessel maintains low neutron flux levels at the primary loop components; the reactor vessel, primary loop components, and auxiliaries are located within a structure (Fig. 1-2) that can contain the steam and fission products released in the highly unlikely event of a large break in the primary piping.[3] The primary system coolant in a PWR is completely isolated from the turbine; this is referred to as an "indirect cycle." In the boiling water reactor

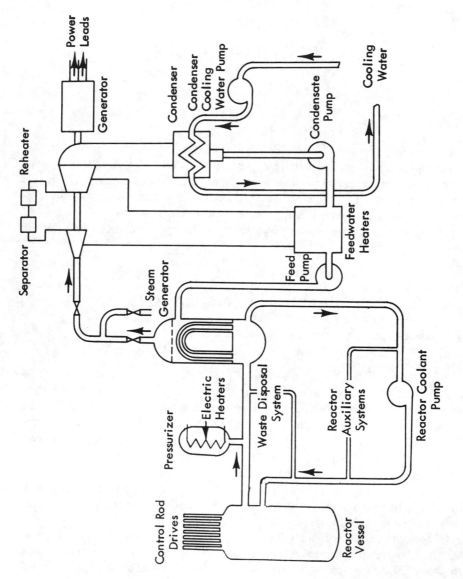

Fig. 1-1.   Typical PWR system (from Ref. 1).

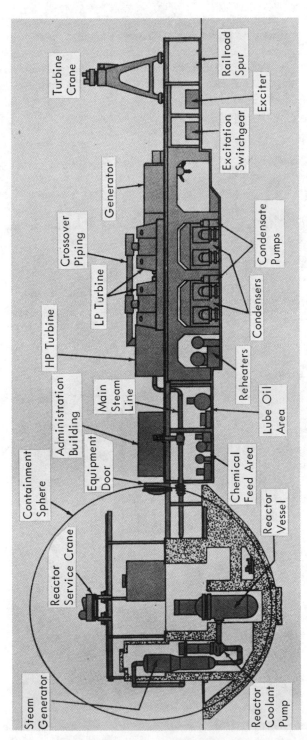

Fig. 1-2. System arrangement for large PWR power plant. [From *Westinghouse Engineer*, **25**, *5*, 145 (1965).]

(BWR), the steam produced in the core goes directly to the turbine; this is called a "direct cycle." Thus, the indirect cycle has the advantage of providing a radio-actively clean secondary system, but requires expensive heat exchangers. Such factors as lower contained volume and simpler control requirements of the PWR compensate for this and make the system economically competitive.

1-1.1.2 Core Configuration

To date, fuel elements for central station power plants have consisted almost entirely of partly enriched uranium dioxide encapsulated in metal tubes. The tubes are assembled into bundles and cooled by water flowing parallel to their axes. Figure 1-3 shows a horizontal cross section of a typical reactor.[4] The fuel assem-blies are placed in a configuration approximating a right circular cylinder; the larger the reactor, the more closely a circular cross section is approached. The

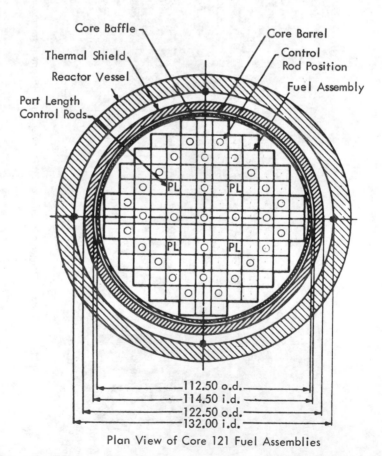

Plan View of Core 121 Fuel Assemblies

Fig. 1-3.   Cross section of the reactor vessel. [From *Proc. Am. Power Conf.*, **30**, 298 (1968).]

assemblies are retained in position by a core baffle attached to the core barrel, which is both the structural support for the baffle and the core support plate on which the fuel assemblies rest. The core barrel is surrounded by a thicker ring of metal designated as the thermal shield. The shield and water gaps between the core and vessel thermalize and attenuate the fast neutron flux emanating from the core. In addition, the shield attenuates the gamma flux; this serves to keep thermal stresses, due to gamma and neutron heating, and total neutron exposure of the reactor vessel within acceptable limits.

The fuel rods are assembled into large square arrays, which are held in position by spring clips on egg-crate grids spaced about two feet apart along the length of the assembly (Fig. 1-4). Fuel tubes are omitted in a number of locations and are replaced by hollow guide tubes that provide the structural support needed to tie the assembly together. The top and bottom nozzles are attached to these guide tubes.

The control rods are clusters of absorber rods that fit into the hollow guide tube. Each control rod member of a cluster consists of a sealed stainless-steel tube filled with a neutron absorbing material such as boron carbide or a silver-indium-cadmium alloy. All fuel assemblies are identical and, hence, each has hollow guide tubes, although only a fraction of the assemblies are under control rod positions. The guide tubes that do not contain control rods are fitted with plugging clusters to prevent unnecessary coolant bypass flow through the empty tubes.

In earlier designs, cruciform-shaped control rods, fitted into cutouts at the edges of the fuel assemblies, were used. To prevent large water gaps when the control rods were withdrawn, the rods were fitted with nonabsorbing extensions. When the control rods were inserted, the extensions projected below the core, necessitating a longer reactor vessel. The individual water slots left by the newer cluster design are small and rod extensions are not needed.

The vertical reactor vessel cross section (Fig. 1-5) illustrates the structural arrangements and coolant flow path.[5] The core barrel is attached to the lower core support barrel, which is hung from a ledge near the top of the vessel. The upper core support plate and upper control rod shrouds are attached to the upper core support barrel, which fits closely within the lowest support barrel and is supported by it. Both upper and lower core support barrels are designed so they can be removed. Normally, the lower support barrel remains in place throughout plant life.

The primary coolant enters through the inlet nozzle and flows downward through the passages between the core barrel and reactor vessel. After turning in the plenum area at the bottom of the pressure vessel, it flows upward through the core and exits through outlet nozzles via close fitting adapters provided as part of the lower core support barrel.

1-1.1.3 Steam Generation

Heat transferred to the secondary fluid is used to generate steam. Consequently, the heat exchanger where this occurs is called a steam generator. A number

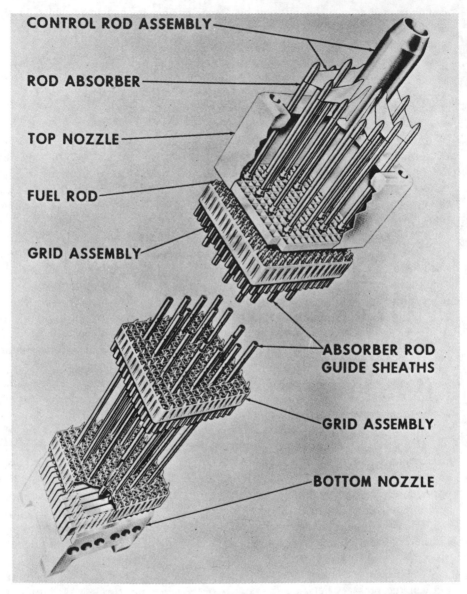

CONTROL ROD ASSEMBLY

ROD ABSORBER

TOP NOZZLE

FUEL ROD

GRID ASSEMBLY

ABSORBER ROD
GUIDE SHEATHS

GRID ASSEMBLY

BOTTOM NOZZLE

Fig. 1-4.　Fuel assembly structure. [From *Proc. Am. Power Conf.*, **30**, 298 (1968).]

of steam generator configurations have been used and, in all, the primary coolant has been circulated through the tubes and the secondary fluid contained within the shell. The low corrosion rates required to limit primary loop contamination led to the use of high-nickel alloys for the tubes. Initially, 300-series stainless steel was used exclusively for this purpose, but its sensitivity to stress corrosion led to the

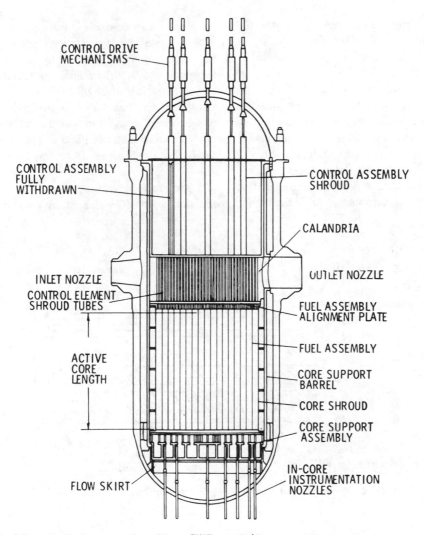

Fig. 1-5.   Vertical cross section of large PWR vessel. (Courtesy of Combustion Engineering.)

use of Inconel in many more recent units. Excessive corrosion rates have also been encountered with Inconel under some conditions. The difficulty appears to be overcome by replacing the phosphate water treatment, originally used for the secondary system, by all-volatile chemistry (ammonia for pH control, hydrazine for oxygen scavenging). However, in some units that operated with phosphate chemistry for an appreciable time, it was found that some corrosion difficulties persisted even after changing to all-volatile chemistry.[a]

---

[a] Fletcher and Malinowski[6] review typical steam generator operating experience.

Horizontal steam generator units are closest to non-nuclear boiler practice. Single drum or reboiler units (Fig. 1-6), where steam separation takes place in the same drum as the bundle, can be used. However, capacity is limited by the maximum drum size that can be fabricated or shipped and by the achievable circulation ratio. Higher capacity can be obtained by using multiple drum units such as shown in Figs. 1-7 and 1-8. Here, steam is generated in the lower drum containing the tube bundle and a two-phase mixture flows upward through the risers to the upper drum where the steam is separated. Primary separation takes place at the vortex created in the separators at the entrance to the drum; water is returned to the lower drum via the downcomers.

Unitized vertical steam generators were originally developed to provide compact, high-efficiency units for marine propulsion systems. Nuclear steam generators

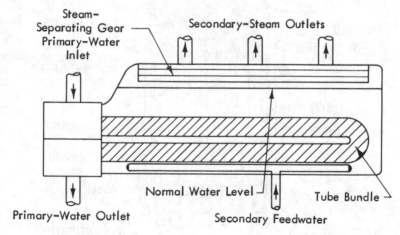

Fig. 1-6.   Horizontal U-tube reboiler. [From *Nucleonics*, **19**, *7*, 71 (1961).]

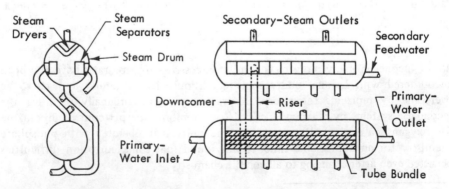

Fig. 1-7.   Horizontal straight-tube multiple drum. [From *Nucleonics*, **19**, *7*, 71 (1961).]

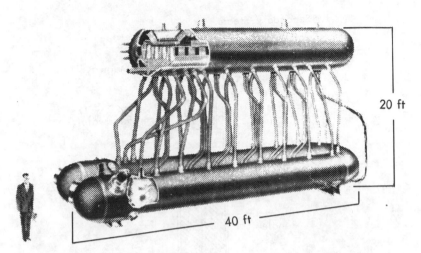

20 ft

40 ft

Fig. 1-8.   Horizontal U-tube with contour shell and multiple drum. [From *Nucleonics*, **19**, *7*, 71 (1961).]

of this type account for the majority of units installed in large central station plants through 1978. All these units have used a U-tube bundle configuration.

A cross-sectional view of a typical vertical natural recirculation unit is shown in Fig. 1-9. The two-phase mixture produced in the tube bundle rises to the primary cyclone separator where the helical propellor establishes a centrifugal field that produces a free vortex. Wet steam exits through the central port and enters the secondary steam separator in the steam drum, while the separated water flows down through the annular passage around the cyclone separator. Feedwater is added through a ring just above the tube bundle; the subcooled water flows downward in the space between the tube bundle and shell, completing the recirculation loop.

Some of the later model steam generators are equipped with an integral preheater located on the cold leg of the tube bundle. Such a preheater increases the secondary surface temperature in the boiling region and produces higher pressure steam. Feedwater is introduced into the preheater through a nozzle in the lower shell. The feedwater flows through a specially baffled region before it joins the main recirculating flow.

Steam generators for Canadian deuterium-uranium (CANDU) reactors are vertical U-tube units generally similar to the units used for vessel PWRs. In most cases, the steam separators are located in the upper section of the vertical unit. However, in the Bruce A generating stations, the outlets of two vertical U-tube steam generators are connected to a single horizontal separating drum.

The desire to produce superheated steam has led to the development of the "once-through" design. A sectional side view of such a unit[7] is shown in Fig. 1-10. To provide a comparison of relative size, the outline of a vertical natural-circulation boiler designed for the same conditions is superimposed on it. The reactor coolant

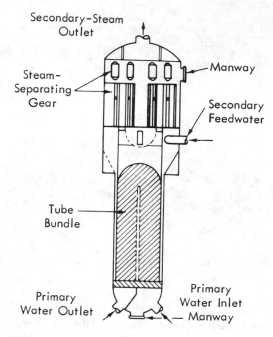

Fig. 1-9.   Vertical U-tube single drum. [From *Nucleonics*, **19**, *7*, 7 (1961).]

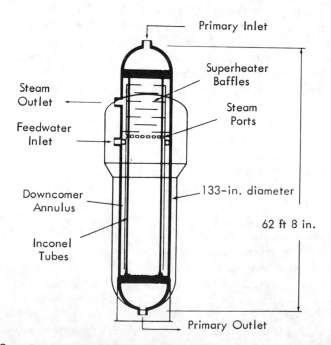

Fig. 1-10.   Once-through steam generator; recirculating unit shown in outline (from Ref. 7).

enters the once-through generator at the upper plenum, flows through Inconel tubes, and exits from the lower plenum. The entering feedwater is sprayed from an annular ring and mixed with steam bled from the upper end of the boiler section. The mixture flows down the annular downcomer and into the tube bank. No distinct water level is present since the mixture quality gradually increases to 100% and superheating starts. The superheater region is baffled to obtain the cross flow and high velocities needed for efficient heat transfer to steam. The superheated steam exits through outlet nozzles near the upper end of the shell.

1-1.1.4 The Saturated Steam Turbine Cycle

Large, modern, fossil-fueled steam plants produce steam at pressures above 3200 psig and temperatures around 1000 to 1050°F. Steam conditions for PWR power plants are vastly different; the relatively low temperature of the primary system coolant, coupled with the necessity of providing an appreciable temperature difference across the heat exchangers, leads to steam pressures in the range of 400 to 950 psig (newer plants are at the upper end of the range).

The majority of PWR plants produce dry and saturated steam. The main drawback of a saturated steam cycle is its excessive endpoint moisture (20 to 24%), which leads to blade erosion and blade efficiency losses that reduce thermal efficiency. Therefore, some method of moisture removal must be used. Evaporation by constant temperature throttling reduces the steam pressure and is detrimental to cycle efficiency. However, the mechanical steam separator has proved to be efficient and desirable since such moisture separators both reduce end-point moisture and improve turbine efficiency.

Efficiency improvement is enhanced when moisture separation is combined with reheating. Figure 1-11 shows the general arrangement of the elements of a power generation cycle for a large PWR using a combination moisture separator/ live-steam reheater, which is placed in the crossover piping between the high- and low-pressure turbines. By means of a finned tube heat exchanger, this device removes moisture and transfers heat from steam bypassed ahead of the throttle valve to the main stream.

Since superheat increases the inlet temperature of a steam turbine and reduces end-point moisture, it improves cycle efficiency considerably. In a closed cycle, superheating can be achieved by using either fossil fuel or the once-through steam generator previously described. The secondary steam from two early demonstration power reactors, the Carolina-Virginias Test Reactor (CVTR) and Indian Point 1, was superheated by fossil fuel. However, the economics of fossil-fired superheat has not proved attractive and the more modern plants have used saturated steam or superheated steam from a nuclear steam generator. Since the primary heat source is at a relatively low temperature, the amount of superheat that can be supplied by such a steam generator is low; i.e., on the order of 50°F. Thus, the steam cycle followed is quite similar to the dry and saturated cycle previously described. Moisture separation or moisture separation plus reheat is still required to prevent excessive end-point moisture.

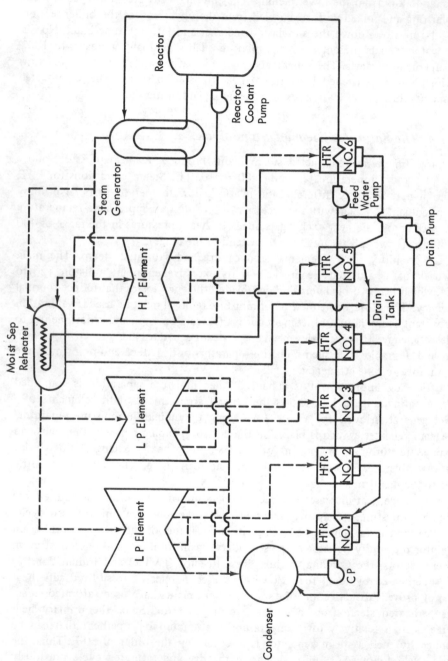

Fig. 1-11.   The PWR steam cycle.

An extensive discussion of thermodynamic cycles for nuclear power plants is provided by Kalafati[8]; conventional equipment is described by Artusa.[9]

1-1.2 Other Reactor Concepts

1-1.2.1 Compact, Light-Water-Moderated Reactor Systems

A compact, low-weight system is of paramount importance in plants designed for ship propulsion. One method[10] achieves this by enclosing the reactor core and steam generator within the same pressure vessel. Coolant flows up through the core and chimney and down through the heat exchangers; it exits through the outer annulus of the double pipes leading to the pumps. The pumps are closely coupled to the reactor, but outside the primary vessel for ease of maintenance. A pressurizer is eliminated by simply maintaining a steam space above the heat exchangers. Since this steam is in equilibrium with water at the average core exit temperature, the quality of the fluid at the hot channel exit is significant.

A compact reactor design of this general type has operated successfully in the merchant ship, *Otto Hahn*. Reactors of this type have also been proposed for combined desalination and power production applications.

In meeting the various military requirements, such as those for small plants in remote areas, weight and size can be of even greater importance. Here, highly enriched fuel can be used. Typical of this reactor group is the PM series of power plants that can be transported by air. The fuel elements are $\sim\frac{1}{2}$-in.-diam hollow tubes made up of a highly enriched UO_2-stainless dispersion between stainless-steel claddings. The fuel is cooled by water flowing both inside and outside the fuel tubes. In PM-1 (Ref. 11), the fuel tubes are assembled into six pie-shaped assemblies and one small central assembly, which are held together by brazed ferrules and end plates.

1-1.2.2 Heavy-Water-Moderated Reactors

Although natural uranium alone cannot bring a light-water-moderated reactor to criticality, it is a satisfactory fuel for large reactors moderated and cooled by heavy water. In areas such as Canada, where enrichment facilities are lacking but heavy water and natural uranium are available, attention has turned to this type of reactor.

The simplest design has the moderator and coolant in the same circuit and is similar in many respects to a light water reactor. However, the greater distance (i.e., the greater slowing-down length) required to thermalize fission neutrons in heavy water requires a much greater water-to-fuel ratio. One proposed method would achieve this by using fuel elements consisting of concentric rings with wide spacing between the rings and fuel elements. Alternatively, clusters of cylindrical fuel rods with wide spacing between clusters can be used. The large size cores that result lead to reactor vessels far larger than those required for a light water plant of the same rating. For large size plants, vessel sizes and weights are beyond present capabilities. Consequently, this concept appears to be practical only for small and medium size plants. The 65-MW(th) Agesta reactor in Sweden[12] is an example of the fuel cluster design.

Separating the moderating and cooling functions helps circumvent the difficulties imposed by the greater slowing down length of heavy water. The fuel elements can be grouped into small clusters, each of which is surrounded by a tube that confines the high-pressure coolant. The pressure tubes are spaced widely to provide enough distance for neutron moderation between them and, with appropriate insulation of the pressure tubes, the moderator outside the tubes can be maintained at low temperatures. Hence, the moderator can be contained in a large vessel, or calandria, which is kept at or near atmospheric pressure. Such thin-walled calandrias can be built in the sizes required for large reactors. Control rods can be placed in the spaces between pressure tubes so rod drive mechanisms operating at high pressures are not required.

To maintain good neutron economy, the pressure tube material must have a low-neutron-absorption cross section. All the reactors of this type have used zirconium-alloy pressure tubes that meet this requirement, while providing a high-strength corrosion-resistant material. In CANDU designs, the pressure tubes are placed horizontally and are of the once-through type.[13] In the CVTR (Ref. 14), vertical U-tubes were used so all connections were made at the calandria top.

The CANDU reactors represent the only heavy-water-moderated PWR design of continuing industrial importance. All current (1978) reactors of this type are fueled with UO_2 pellets produced from natural uranium. Reactors fueled by thoria slightly enriched by ^{233}U have been proposed.[15] If fabrication and reprocessing losses can be held low, a self-sufficient fuel cycle can be possible. That is, the bred ^{233}U just may be sufficient to replace the ^{233}U consumed in the fission process.

1-1.2.3 Graphite-Moderated Reactors

Graphite-moderated and light-water-cooled reactors have a long history in the production of plutonium. It is natural that this concept be adapted to power generation by use of higher coolant temperatures. All such reactors built to date (1978) have maintained an overpressure sufficient to prevent bulk boiling. Hence, they fit our definition of a PWR. The first atomic power station in the USSR was of this nature. The reactor is encased in a sealed cylindrical steel jacket filled with graphite bricks pierced by fuel channels; an atmosphere of helium or nitrogen is maintained to prevent graphite oxidation. The fuel element design[16] was rather unusual: The active section of each fuel channel consisted of a graphite cylinder pierced by five tubes. The coolant flowed down the central tube, then upward through passages inside the annular fuel elements. The fuel elements consisted of slightly enriched hollow uranium-metal cylinders clad on both sides by thin stainless steel. The complexity of the fuel element design makes this reactor ill suited for large central-station application.

A large, dual-purpose reactor (power and plutonium production) was erected at what was then the Hanford laboratory of the U.S. Atomic Energy Commission (AEC). The reactor, designated the NPR, consisted of a very large, 33- X 33- X

39-ft rectangular stack of graphite, penetrated by ~1000 horizontal pressure tubes.[17] The 2.7-in.-i.d. Zircaloy pressure tubes contained annular fuel elements of uranium metal clad in Zircaloy. If the reactor were converted to power production alone, UO_2 could replace the uranium metal.

Heat was removed from the NPR by six loops, each containing two steam generators. During dual-purpose operation, a portion of the low-pressure steam proceeded to two turbines, while the remainder proceeded to 16 dump condensers. For power production only, the dump condensers could be eliminated and higher pressure steam could be generated.

1-1.2.4 Supercritical Reactors

Reactors using supercritical water substance, or water above the critical pressure, as a reactor coolant, have been proposed for future applications. Such systems have the potential advantages of higher efficiency due to higher steam pressure and temperature, higher enthalpy rise through the core, avoidance of limitations imposed by boiling heat transfer, and lower contained energy in the system. Both direct and indirect cycle reactors were considered. Since no phase change can occur at supercritical pressures, both cycles fall under our definition of a PWR.

Supercritical reactors can be of the pressure tube or pressure vessel type. In the SCOTT-R system,[18] the reactor is moderated by heavy water (or graphite) and cooled with light water at supercritical pressure. Each pressure tube contains a fuel assembly of slightly enriched UO_2 in the form of annular rings. A direct cycle, once-through coolant system is used where all the reactor coolant flows directly to the turbine, which eliminates all recirculating equipment in the primary path.

Several pressure vessel concepts have been advanced, the simplest of which is a supercritical indirect-cycle water reactor. In this concept, an indirect-cycle, open lattice, PWR is operated at supercritical pressures.

Supercritical reactors require the successful development of a high-temperature collapsed-cladding fuel element; cladding is so thin that it relies on the fuel for its support. In addition, water chemistry conditions that will suppress radiolysis, control corrosion, and prevent deposition of solids on the fuel must be determined. There is little current (1979) interest in such systems.

1-1.2.5 Status of Various Concepts

Although a variety of pressurized water concepts for generation of central station power have been examined, most of these are no longer actively considered. For large central stations, only two design types have survived: (a) the CANDU heavy-water-moderated design and (b) the vessel-type light-water-moderated system at subcritical pressure (basic concept). The CANDU design is used in Canada as well as abroad. However, the vessel-type design, using slightly enriched fuel, has dominated PWR plants constructed in the U.S., Western Europe, and Japan. The USSR also appears to have abandoned its graphite-moderated concept. Later PWR plants built by the USSR are of a vessel design generally similar to that used in the

U.S. Since the basic vessel-type reactor dominates PWR practice, this monograph concentrates on its analysis.

1-2 POWER GENERATION AND DISTRIBUTION IN VESSEL-TYPE REACTOR CORES

1-2.1 Power Distribution in Unperturbed, Uniformly Loaded Cores

Since light-water-moderated reactors cannot go critical using entirely natural uranium fuel, the designer has three choices:

1. to utilize fuel elements that all contain slightly enriched fuel

2. to provide a "spiked core," where highly enriched fuel elements are uniformly dispersed in a matrix of natural uranium elements

3. to provide a region of highly enriched fuel surrounded by low enrichment or natural uranium fuel.

The last alternative forms the basis of the seed-blanket design. Dispersion of enriched elements among natural uranium fuel results in power outputs for the enriched rods substantially greater than for the others. It is difficult to provide additional flow to the enriched elements and they could limit power output. Further, significantly lower fuel costs can be obtained by using slightly enriched fuel in all elements. This is the common practice for large, central station power plants. No practical reactor design has used a spiked core.

If the fuel elements were dispersed through the core in a uniform manner and if fuel enrichment were uniform throughout, we can roughly approximate overall neutron behavior by considering the core to be homogeneous. Most light water designs are essentially unreflected. Thus, we consider the unperturbed core as a bare homogeneous reactor. For such a core, an approximation of the neutron flux distribution can be obtained by solving the wave equation[19]

$$\nabla^2 \phi + B^2 \phi = 0 \ , \tag{1.1}$$

where ϕ is thermal neutron flux and B^2 is geometric buckling. Our boundary conditions for the solution of this equation are that flux goes to zero at the extrapolated boundaries of the reactor and that it is finite everywhere except at the extrapolated boundaries. The extrapolated boundaries are approximated by adding 0.71 λ_{tr}, where λ_{tr} is the transport mean-free-path (mfp) of a neutron, to the distance from the center line. Thus, for a cylindrical reactor, extrapolated length L' is

$$L' = L + 1.42 \, \lambda_{tr} \ , \tag{1.2}$$

where L is the actual length of the reactor and R', the extrapolated radius, is given by

$$R' = R + 0.71 \, \lambda_{tr} \ , \tag{1.3}$$

where R is the actual radius. For large reactors, λ_{tr} is negligible with respect to R and L and can be ignored. The solution for several of the most common geometries is listed in Table 1-I (Ref. 20). Under the conditions we have postulated, and if the effects of burnup are not significant, power generation is proportional to thermal neutron flux. Thus, the functions indicated in Table 1-I provide an overall description of power distribution.

TABLE 1-I

THERMAL FLUX DISTRIBUTION IN BARE HOMOGENEOUS REACTORS

Geometry	Coordinate	Distribution Function
Infinite slab	x	$\cos\left(\dfrac{\pi x}{L'}\right)$
Rectangular parallelepiped	$x, y,$ or z	$\cos\left(\dfrac{\pi x}{L'}\right), \cos\left(\dfrac{\pi y}{L'}\right), \cos\left(\dfrac{\pi z}{L'}\right)$
Sphere	r	$\dfrac{\sin(\pi r/R')}{\pi r/R'}$
Finite cylinder	$\begin{cases} r \\ z \end{cases}$	$J_0(2.405\, r/R')$ $\cos\left(\dfrac{\pi z}{L'}\right)$

Power distribution within a reactor free of control rods is not only of academic interest; two control systems lead to such a situation for base-loaded plants. In the chemical shim control system, boric acid is added to the reactor to compensate for excess reactivity at the beginning-of-life (BOL). However, it is gradually withdrawn as core reactivity declines with lifetime. A similar procedure is used in spectral shift control,[b] where the ratio of D_2O to H_2O in the moderator is decreased in response to the change in reactivity with lifetime. Decreasing the amount of D_2O present increases the number of thermal neutrons present in the core, thus increasing reactivity and compensating for the decrease in reactivity due to fuel burnup. At full power, such reactors can be operated with all control rods essentially withdrawn.

Power distribution at BOL, in the uniform, unrodded cylindrical reactor, is closely approximated by a J_0 radial distribution and an axial cosine distribution. Thus, we write

$$q''' = J_0\left(2.405\, r/R'\right)\cos\left(\frac{\pi z}{L'}\right) q'''_{max} , \tag{1.4}$$

---

[b]This control system is no longer of commercial interest.

where

q''' = rate of volumetric heat generation

q'''_{max} = maximum volumetric heat generation at the center of the core

r = radial location

z = axial location (from centerline).

1-2.2 Effect of Fuel Loading

Most of the early power PWRs were uniformly loaded. In some small plants, such as the portable military reactors of the PM series, where simplicity of refueling is important, this loading scheme can still be followed. For large, central station plants, one disadvantage is that it leads to high power peaking in the central region of the core. In addition, relatively low-average burnups are obtained since even at the end of life (EOL) the outermost fuel elements have been only moderately burned down. One means of alleviating these difficulties is to use a zone loading scheme, where the core is loaded with fuel elements of three or more different enrichments. The most highly enriched fuel elements are placed in the outermost position; the least enriched are placed in the central region. Since power is approximately proportional to the product of thermal neutron flux and fissionable fuel concentration, the power level in the inner region of the core is lowered and that of the outer region is raised. This power flattening can significantly increase the total power capability of the core. When criticality is no longer possible, the entire core is replaced.

Replacement of the complete core at each refueling leads to a low-average core burnup. This is increased in most fuel cycling schemes by replacing only a portion of the fuel at each refueling. In the simplest of these schemes, the initial core is zone loaded as described above. At the end of the first core life, the most burned fuel from the central core region is removed, the outer regions are moved inward, and fresh fuel is placed in the vacated spaces in the outermost region. The same procedure is followed in all subsequent core loading so that at equilibrium all the removed fuel has been through three or more cycles and has a high burnup.

Several variations of this fuel cycling scheme are available. One significant departure is the so-called "Roundelay" method.[21] In the equilibrium cycle, fresh elements are inserted in uniform distribution throughout the core at each refueling. For example, with Roundelay 3 batch, every third element is replaced while the other elements are left in place. In Fig. 1-12, the fuel elements are numbered in the order in which they might be replaced in the reference core; i.e., during one refueling all No. 1 fuel assemblies would be replaced, and so on. Mixing fresh and burned assemblies produces a strong coupling between them and permits improved power production from the burned elements.

Figures 1-13 and 1-14 show a typical radial power distribution in equilibrium cores after refueling for three-region and Roundelay-loaded reactors. The power

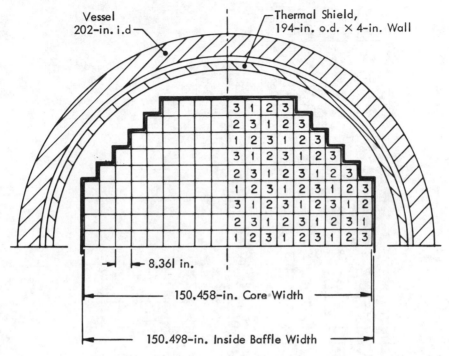

Fig. 1-12.   Roundelay fuel cycling scheme (from Ref. 21).

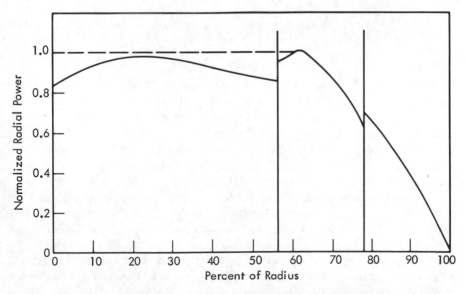

Fig. 1-13.   Normalized radial power distribution in a PWR using three-region batch loading.
[From *Nucl. Eng. Des.*, 6, 301 (1967).]

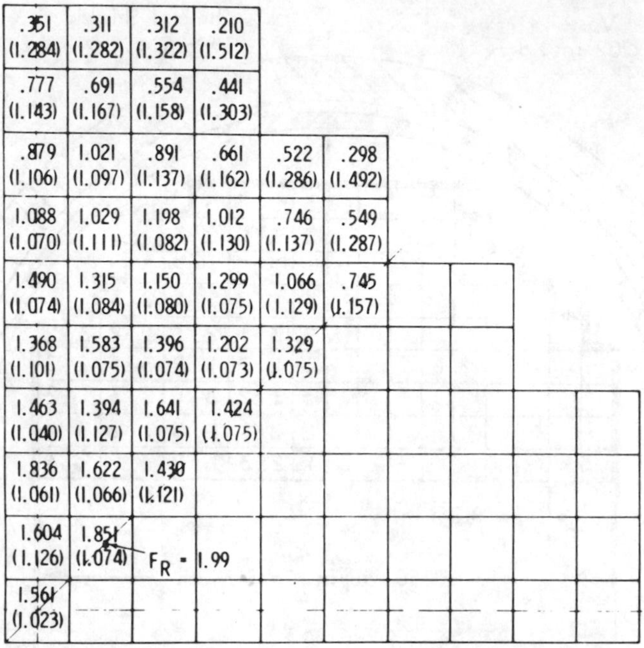

.361	.311	.312	.210		
(1.284)	(1.282)	(1.322)	(1.512)		
.777	.691	.554	.441		
(1.143)	(1.167)	(1.158)	(1.303)		
.879	1.021	.891	.661	.522	.298
(1.106)	(1.097)	(1.137)	(1.162)	(1.286)	(1.492)
1.088	1.029	1.198	1.012	.746	.549
(1.070)	(1.111)	(1.082)	(1.130)	(1.137)	(1.287)
1.490	1.315	1.150	1.299	1.066	.745
(1.074)	(1.084)	(1.080)	(1.075)	(1.129)	(1.157)
1.368	1.583	1.396	1.202	1.329	
(1.101)	(1.075)	(1.074)	(1.073)	(1.075)	
1.463	1.394	1.641	1.424		
(1.040)	(1.127)	(1.075)	(1.075)		
1.836	1.622	1.430			
(1.061)	(1.066)	(1.121)			
1.604	1.851	F_R = 1.99			
(1.126)	(1.074)				
1.561					
(1.023)					

FOR EACH ASSEMBLY

Upper number is (average power in assembly)/(average power in core).

Lower number is (maximum power in assembly)/(average power in assembly).

Fig. 1-14. Typical radial power distribution in Roundelay-loaded reactor (from Ref. 21).

distributions shown are the ratio of the power at the given point to the average power across the core at that axial location. The marked "ripple" in the Roundelay power distribution results from significantly different power production from adjacent fuel assemblies with different burnups. In general, for both refueling schemes, as burnup proceeds, peak power decreases from that shown and power in the lower power region increases. Note that small changes in fuel assembly placement can significantly affect peak power.

In actual practice, one-fourth to one-third of the core is generally reloaded at a given time, but the fuel placement scheme followed is considerably more sophisticated than either the three-region or Roundelay procedure. A detailed study of possible placement schemes is made; the "scatter-loading" scheme, which provides the best economics, is selected. In view of the very large number of placement policies that can be followed, a methodical procedure limiting the interchanges considered is required. Rothleder[22] lists a series of heuristic rules that can be used to limit the number of policies considered. Other investigators[23-25] proposed using simplified procedures for estimating the radial power generation coupled with mathematical programming techniques for minimalization of power costs.

As noted in Sec. 1-2.1, the nuclear designer can choose a "seed and blanket" design as an alternative to using slightly enriched uranium throughout the core. Such a core consists of one or more relatively small seed regions containing highly enriched fuel assemblies, and blanket regions of natural or low-enrichment fuel surrounding the seed regions. The blanket is composed of material with a multiplication factor less than unity and cannot be made critical by itself. Reactivity of the core is primarily determined by the properties of the seed; thus, control rods are required only in the seed.

The geometry initially considered was that of a cylindrical annular seed with blanket regions both inside and outside. This arrangement (see Fig. 1-15 and Refs. 26 and 27) was used in Shippingport Core I. The blanket elements were bundles of zirconium-alloy-clad cylindrical rods containing UO_2 pellets fabricated

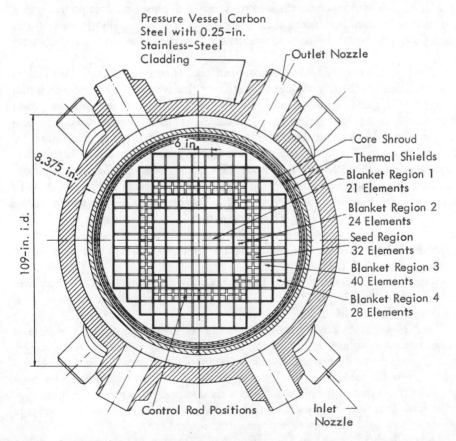

Fig. 1-15.   Cross section of Shippingport Reactor, Core 1. [From *Directory of Nuclear Reactors*, Vol. IV, *Power Reactors*, pp. 21-32 (1962).]

from natural uranium. The seed elements consisted of a series of fuel plates held by Zircaloy alloy boxes, each segmented into four quadrants. The space between the quadrants was used for a control rod passage. The fuel plates themselves consisted of an enriched fuel alloy strip sandwiched between two zirconium alloy cover plates and four side strips. In later seeds, a dispersion of UO_2 in Zircaloy replaced the U-Zr alloy fuel.

The power distribution within a seed and blanket reactor is appreciably different from the distribution heretofore considered. In reactors of this type, infinite multiplication of the annular seed region is greater than unity, while that of the blanket is less than one. The higher fuel content and multiplication factor of the seed result in a higher fast neutron flux in the seed. In addition, the highly enriched fuel in the seed has a higher thermal absorption cross section than the blanket. This results in a leakage of fast neutrons from the seed into the blanket and a leakage of thermal neutrons in the opposite direction.

In the blanket, ^{238}U fast fission produces the power at BOL. As burnup proceeds, formation of plutonium in the blanket creates new power-producing fuel. Leakage of fast neutrons from the seed thus results in power production in a region that cannot by itself maintain a chain reaction. It is economically desirable for as much power as possible to be produced in the relatively low-cost natural uranium blanket.

The power division between blanket and seed is one of the important factors in determining the thermal design of a reactor of this type. The one-group diffusion theory in Eq. (1.1) is not adequate to obtain realistic estimates of the power distribution. If a simple geometry of concentric radial rings is assumed, it is possible, in principle, to use two-group diffusion theory to make such calculations by hand using a modification of the matrix methods described in Ref. 28. However, the calculations are tedious, and because there are shifts in the power distributions over life, the calculations must be repeated at several points during the cycle to carry out a realistic analysis of the design. Furthermore, determination of the nuclear parameters invariably involves a depletion calculation that depends on the past history of the power distribution. The combined result of these complications is that in practice seed-blanket power distribution analysis is now carried out by computer.

Radkowsky and Bayard[29] discuss the procedures used for calculating the power distribution in Shippingport Core I. At BOL, the ratio of blanket power to total power was calculated as ~0.52.

Seed power does not decrease as rapidly as seed volume and, therefore, a decrease in seed volume results in increasing seed power density. Hence, power sharing tends to be determined by the maximum power density that can be extracted from the seed. Thus, the average power density in the seed can be more than four times that of the blanket.[29] Since resonance capture is of no consequence in the seed, it is feasible to use thin fuel elements with a high surface area per core volume to attain the high power removal rates required.

On the basis of data from Shippingport, the fraction of total power produced in the blanket increases during the lifetime of any one seed. However, blanket power drops on the insertion of each successive seed. It dropped from 58% for the first seed, to 48% in the average of the fifth seed, and to 44% by the tenth seed.

Due to the short radial dimension of the seed, the radial peak-to-average power ratio in the seed is not far from unity. However, radial power distribution in the blanket is poor because the neutron flux falls off markedly with distance from the seed. A typical radial power distribution, illustrating peaking adjacent to an annular seed, is shown in Fig. 1-16 (Ref. 30). In the example shown, peak seed-to-average seed power is 1.15, while the peak-to-average blanket power is 1.85. During the operating life of a seed, the power distribution is relatively stable and, hence, it is feasible to orifice coolant flow to the various positions of the blanket.

Power peaking in blanket elements adjacent to the seed can be reduced somewhat by the design used in Shippingport Core 2 (Ref. 26). Here seed assemblies alternate between two fuel assembly rings; however, peaking is still considerable

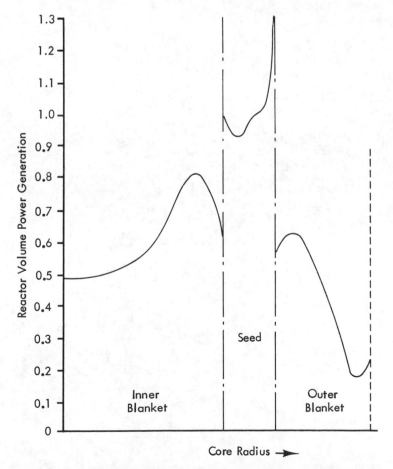

Fig. 1-16. Power distribution in a seed and blanket reactor (after Ref. 31).

and two-pass coolant flow is provided to improve performance. Blanket fuel element configuration was also changed. In Core 2, the blanket fuel elements consist of UO_2 platelets inserted between slotted Zircaloy plates that are welded together; the plates are then assembled into a rectangular box. Fuel elements of this design are capable of withstanding burnups considerably in excess of those that can be provided by cylindrical fuel rods.

Later studies with the seed and blanket concept have been directed toward developing a light-water breeder reactor (LWBR). The concept is based on the ^{232}Th-^{233}U cycle. Both the seed and blanket consist of ThO_2 fuel rods enriched with varying amounts of $^{233}UO_2$. (Seed rods contain up to 6 wt% UO_2, while blanket rods contain up to 3 wt% UO_2.) From the cross section of the first Shippingport LWBR reactor [Fig. 1-17 (Ref. 31)], we see that the core consists of a symmetrical array of hexagonal modules surrounded by a reflector blanket. Each core module contains an axially movable seed region ($k > 1$) and a stationary annular blanket ($k < 1$). Both seed and blanket regions are made up of arrays of tightly packed cylindrical fuel rods containing oxide pellets.

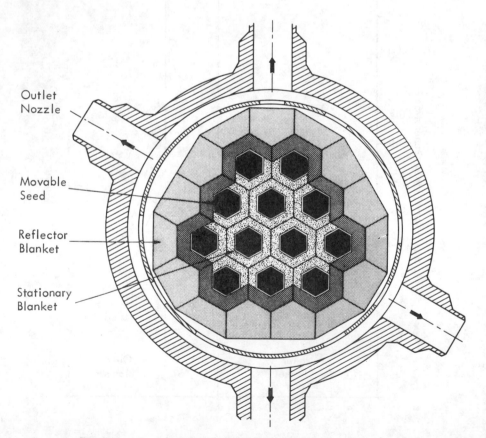

Outlet
Nozzle

Movable
Seed

Reflector
Blanket

Stationary
Blanket

Fig. 1-17.   Cross-sectional view of the first Shippingport LWBR core.

Control of the LWBR is accomplished by moving the seed. As the seed elements are moved out of core, leakage is increased and the core becomes subcritical. This control procedure, which avoids losing neutrons to poisons, is necessary since the calculated breeding ratio is only slightly in excess of unity.

1-2.3 Effect of Control Rods, Water Slots, and Voids

1-2.3.1 Effects of Control Rods on Power Distribution

In some reactor designs, all excess reactivity is compensated for by inserting control rods. The presence of the strongly absorbing material perturbs the flux both axially and radially. However, this is not entirely a disadvantage since the rods can be used to improve radial flux distribution. For example, some, or all, of the rods in the center of the core can be partially inserted at BOL, thereby reducing flux peaking. This effect is illustrated in Fig. 1-18, where we observe that flux essentially goes to zero at the boundaries of the control rods, as well as at the extrapolated boundary of the reactor. The flux, and hence, the power level in the outer region are increased concurrently with the decrease in the central region.

The improvement in radial distribution attainable by inserting rods is often more than negated by the change in axial distribution. The usual situation in a PWR where the rods are top entering is shown in Fig. 1-19. At BOL, the partially inserted rods skew the flux toward the bottom. As the rods are withdrawn, the less-burned fuel at the top of the core causes the flux to be skewed toward the top at EOL. As observed in Fig. 1-19, the ratio peak-to-average power can be higher than that of the unperturbed core.

Power distribution of the type shown in Fig. 1-19 generally can be represented by modifying the usual cosine distribution to the form

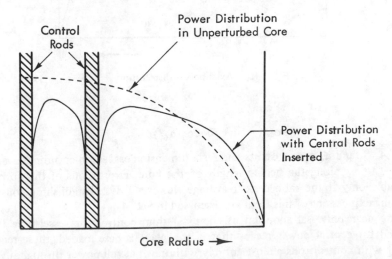

Fig. 1-18.   Radial power distribution in a cylindrical reactor with and without control rods.

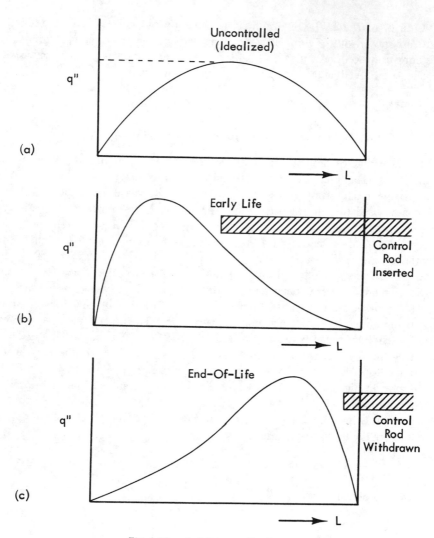

Fig. 1-19. Axial power distribution.

$$\phi(l) = (A + Bl)\cos(\alpha_0 l) \ , \tag{1.5}$$

where A, B, and α_0 are constants and l is a dimensionless length coordinate varying from -1 to $+1$ along the heated length of the core. Expressions of this form are applicable only if the ratio peak-to-average flux is <1.892. Establishing constants and using expressions of this form are discussed in Sec. 4-1.1.

The near universal adoption of chemical shim control for vessel-type PWRs designed for central power means that if the plant is base loaded, these reactors operate with control rods almost entirely withdrawn at full power throughout core life. This minimizes axial flux distortion due to rod motion.

1-2.3.2 Power Perturbations Due to Structural Material, Water Slots, and Voids

No fuel-water lattice is entirely uniform; additional structural material is present at all grid locations, in fuel assembly cans, in the support structure required at core edges, etc. Since such material is a neutron absorber without being a moderator or source, it causes a local decrease in flux and, hence, in power. The effect can be small if the material is of low cross section (i.e., Zircaloy), but it can be substantial if a high cross-section material, such as stainless steel, is used.

Perhaps of even more concern to the designer is local peaking due to additional water slots in a light-water-moderated reactor. Such spaces result from clearance between fuel assemblies, variations in lattice dimensions, and slots left by the withdrawal of control rods. Since water provides additional moderation, the local thermal flux and, hence, power, increases. This effect is illustrated in Fig. 1-20 for a circular water hole in a low enrichment, stainless-steel-clad core.[32] Slots of a size that could cause substantially more peaking than illustrated can easily arise; care is taken to avoid these. For example, where cruciform or Y-shaped control rods are used, followers of a low cross-section material are provided to prevent large water slots from being introduced when the rods are withdrawn. Such followers are not required with rod cluster control (RCC) and, as previously noted, the RCC concept allows the use of shorter reactor vessels.

Creation of a steam void at the exit of the warmest core region is an additional factor that distorts normal flux patterns. In general, replacement of water moderator by steam reduces core reactivity and, hence, flux and power, in the region of

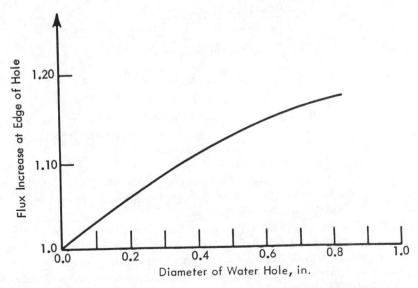

Fig. 1-20. Flux peaking at the edge of a circular water hole. [From *Nucl. Eng. Des.*, 6, 301 (1967).]

the voids. The effect of such voids is particularly significant during transient and accident situations where coolant enthalpy rises appreciably above its normal value. Power reduction due to void creation can decrease the severity of some situations.

1-2.4 Nuclear Hot-Channel Factors

The thermal designer can readily determine average core parameters from a knowledge of total heat output, core heat transfer data, and flow to the core. However, core performance is not limited by average conditions, but by the most severe conditions. It is convenient and useful to define the hot channel of the core as that coolant channel or flow path where core heat flux and enthalpy rise is a maximum. Conditions in the hot channel are defined by several ratios of local-to-average conditions. These ratios are referred to as "hot-channel factors." Those aspects of the complex nuclear data that are most significant to the thermal designer can be conveyed by three nuclear hot-channel factors. We define

$$F_R^N = \text{Radial Nuclear Factor}$$

$$= \frac{\text{Mean Heat Flux in Hot Channel}}{\text{Mean Heat Flux in Average Channel of Core}} \tag{1.6}$$

$$F_Z^N = \text{Axial Nuclear Factor}$$

$$= \frac{\text{Maximum Heat Flux in Hot Channel}}{\text{Mean Heat Flux in Hot Channel}} \tag{1.7}$$

$$F_Q^N = \text{Nuclear Heat Flux Factor}$$

$$= \frac{\text{Maximum Heat Flux in the Core}}{\text{Mean Heat Flux in the Core}} \cdot \tag{1.8}$$

From the above definitions, it immediately follows that

$$F_Q^N = F_Z^N F_R^N \cdot \tag{1.9}$$

Some authors define F_Z^N as the highest value that the ratio (maximum channel flux-to-average flux of given channel) can have and, when so defined, the maximum of the ratio may not occur in the hot channel. When using reported values of F_Z^N, we should determine the basis for their definition.

The above nuclear factors are defined on the basis of nominal channel dimensions. In computing limiting heat fluxes and enthalpy rises, consideration must also be given to deviation from nominal dimensions and flow. The procedure used to account for these deviations is discussed in Sec. 5-1.

An estimate of the magnitude of the nuclear hot-channel factors can be obtained by considering a large, bare, homogeneous reactor of cylindrical shape. For a uniformly enriched, unpoisoned, unperturbed reactor, we see that the heat flux follows a J_0 radial distribution and a cosine axial distribution. Figure 1-21 shows axial and radial nuclear hot-channel factors for such a reactor as a function of the ratios of (actual length)/(actual length + extrapolation length). For large

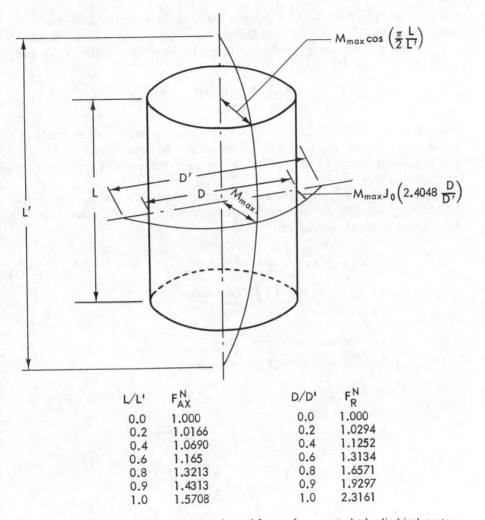

L/L'	F_{AX}^N		D/D'	F_R^N
0.0	1.000		0.0	1.000
0.2	1.0166		0.2	1.0294
0.4	1.0690		0.4	1.1252
0.6	1.165		0.6	1.3134
0.8	1.3213		0.8	1.6571
0.9	1.4313		0.9	1.9297
1.0	1.5708		1.0	2.3161

Fig. 1-21. Axial and radial nuclear hot-channel factors for unperturbed cylindrical reactors.

reactors, the extrapolation length becomes negligible, and the values for L/L' and D/D' approach unity. Thus, we find that under these idealized conditions F_Q^N = 1.5708 × 2.3161 = 3.638. In an actual uniformly loaded core, the overall hot-channel factor would be higher due to such effects as lattice nonuniformities, control rods, changes in fuel and fission product concentration, and coolant property variation. These effects can often be more than compensated for by improvements in F_R^N that can be obtained through nonuniform loading.

For a seed and blanket core, the overall core nuclear hot-channel factor has little meaning in view of markedly different power generations in the seed and blanket. It is useful to extend the concept to define hot-channel factors for each region of the core. We can define separate seed and blanket hot-channel factors by using Eqs. (1.6), (1.7), and (1.8) with the word core replaced by seed or blanket.

The entire radial power distribution in any core can be described by defining a local radial factor, F_{RI}^{N}, such that

$$F_{RI}^{N} = \frac{\text{Mean Heat Flux in Given Channel}}{\text{Mean Heat Flux in Average Channel of Core}} \cdot \qquad (1.10)$$

For a typical core, Fig. 1-22 illustrates the relative number of fuel elements with radial nuclear hot-channel factors in a given range. The peak radial flux is reached in a very small number of elements (often just one); the majority of the elements have radial factors <1.0.

An accurate value of F_{Q}^{N} for a real core requires a three-dimensional nuclear calculation for evaluating power throughout the core. The nuclear heat flux factor can then be calculated from

$$F_{Q}^{N} = \frac{P'(x_0,y_0,z_0)}{(1/V) \iiint P'(x,y,z)\,dx\,dy\,dz} , \qquad (1.11)$$

where

    $P'(x,y,z)$ = local power density at (x,y,z)

    (x_0,y_0,z_0) = location of peak local power density

    V = core volume.

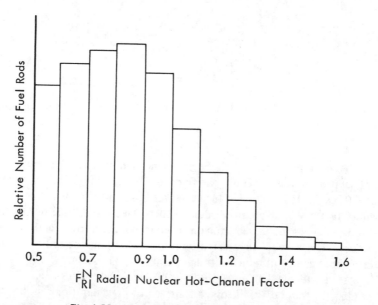

Fig. 1-22.   Power census in a typical rodded core.

Accurately determining the highest possible F_Q^N that could occur during life would require three-dimensional power calculations to be conducted for a very large number of conditions in a load-following plant. To avoid the expense of these calculations, a simpler synthesis procedure is often used.

The synthesis approach defines two additional factors:

$F_{xy}(z)$ = peak-to-average power density at axial elevation z

$$P'(z) = \text{relative core power at axial elevation} \left[\int_0^L P'(z)\,dz = 1 \right].$$

With these definitions, the value of F_Q^N is then given by

$$F_Q^N = \max[P'(z) \cdot F_{xy}(z)] \ . \tag{1.12}$$

In the synthesis approach, the values of $P'(z)$ are obtained from separate calculations. The values of $P'(z)$ are estimated from a one-dimensional axial calculation; values of $F_{xy}(z)$ are obtained by two-dimensional calculations at a number of planes. By appropriate modeling, it is possible to obtain

1. a one-dimensional axial distribution model that is consistent with three-dimensional average axial power distribution behavior during reactor operation

2. two-dimensional values for $F_{xy}(z)$ that are always equal to or greater than the value of $F_{xy}(z)$ obtained for a three-dimensional calculation.[33]

The synthesis approach yields conservative values for F_Q^N under these conditions.

1-2.5 Heat Generation Within Fuel Elements

It is often assumed that the rate of power generation is essentially constant across the fuel rod. While this is often satisfactory as a first approximation, it is not strictly true. As previously noted, heat generation in a volume element of fuel of a thermal reactor is essentially proportional to the thermal neutron flux at that point. Since thermal neutron flux decreases toward the center of the element due to neutron absorption in the fuel, power generation decreases similarly.

Following the explanation in Ref. 34, we consider a unit cell of a uniform rectangular lattice containing cylindrical fuel rods of radius R_0. The problem is simplified by supposing that the square unit cell is replaced by a circle, radius R_1, with the equivalent area (Fig. 1-23). Further, we assume that the neutron slowing down is constant in the moderator and zero in the fuel. The thermal diffusion equation for monoenergetic neutrons in the fuel rod is then

$$D_1 \nabla^2 \phi - \Sigma_a \phi = 0 \ , \tag{1.13}$$

when ϕ is neutron flux at any point in the rod, D_1 is the diffusion coefficient, and Σ_a is the macroscopic absorption cross section. After dividing by D_1 and replacing Σ_a/D_1 by κ_0^2, we obtain

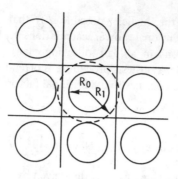

Fig. 1-23.   Equivalent cell of fuel rod lattice.

$$\nabla^2\phi - \kappa_0^2\phi = 0 \ . \tag{1.14}$$

In cylindrical coordinates, this becomes

$$\frac{d\phi^2}{dr^2} + \frac{1}{r}\frac{d\phi}{dr} - \kappa_0^2\phi = 0 \ . \tag{1.15}$$

Since κ_0^2 is positive, Eq. (1.15) is a modified Bessel equation with the general solution

$$\phi = AI_0(\kappa_0 r) + A'K_0(\kappa_0 r) \ , \tag{1.16}$$

where I_0 and K_0 are zero order, modified Bessel functions of the first and second kind, respectively. The second term can be eliminated since K_0 would require the neutron flux to go to infinity at the axis of the fuel rod. Thus, for the flux in the rod we obtain

$$\phi = AI_0(\kappa_0 r) \ . \tag{1.17}$$

The value A can be determined from the boundary condition that ϕ is equal to $\phi_{surface}$ at $r = R_0$.

It is also useful to relate neutron flux at the surface of the fuel rod to average flux in the fuel with the definition

$$F = \frac{\text{Thermal Neutron Flux at Fuel Rod Surface}}{\text{Mean Thermal Flux in Interior of Rod}} \ . \tag{1.18}$$

With the previous assumptions, for a cylindrical rod we can show that

$$F = \frac{\kappa_0 R_0}{2}\frac{I_0(\kappa_0 R_0)}{I_1(\kappa_0 R_0)} \ . \tag{1.19}$$

For an infinite slab fuel element, we obtain

$$\phi = \frac{\phi_0 \cosh(\kappa_0 x)}{\cosh(\kappa_0 a)} \ , \tag{1.20}$$

$$F = \kappa_0\, a \coth(\kappa_0\, a) \ , \tag{1.21}$$

where

    x = distance from the centerline

    a = half thickness of the fuel element

    ϕ_0 = surface flux.

The previous assumption (that power distribution across the fuel rod is directly proportional to thermal flux) is not entirely correct since as burnup proceeds, the fuel and poison distribution are no longer uniform. In view of the usual uncertainties in flux levels and in the physical properties of the fuel elements, consideration of these nonuniformities is rarely required. Perhaps more important are the limitations on simple diffusion theory. More exact calculations show that diffusion theory tends to underestimate flux depression in the fuel.

Most modern computer calculations of flux depression are based on escape probabilities. In particular, the method of Amouyal et al.[35] is widely used. This method is carried out only for the equivalent cell; the neutron current is taken to be zero at the outer surface of this cell. However, diffusion theory is not assumed to hold within the fuel. Amouyal et al. calculate p_F as the probability that neutrons produced uniformly within the fuel ultimately escape from the fuel. The expression obtained for a cylindrical fuel rod of radius R_0 (cm) is

$$\frac{1}{p_F} = 1 + (\Sigma_{aF}/\Sigma_{tF})\{A\,[1 + \alpha(\Sigma_{sF}/\Sigma_{tF}) + \beta(\Sigma_{aF}/\Sigma_{tF})^2] + R_0\Sigma_{tF}\} \ , \tag{1.22}$$

where

$$A = \frac{1 - p_{F_0}}{p_{F_0}} - R_0\Sigma_{tF}$$

p_{F_0} = probability that neutron escapes without collision from an infinite cylinder of radius R and macroscopic cross section Σ_{tF} (given by Fig. 1-24)

Σ_{aF} = macroscopic absorption cross section of the fuel, cm^{-1}

Σ_{tF} = total macroscopic cross section of the fuel, cm^{-1}

Σ_{sF} = macroscopic scattering cross section of the fuel, cm^{-1}

α, β = functions of $R_0\Sigma_{tF}$ plotted in Fig. 1-25.

It may be shown that $1/p_F$ is equivalent to F. Further details on this method are available from Lamarsh.[36]

1-2.6 Distribution of Power Among Fuel, Moderator, and Structure

In addition to axial and radial power distributions, the thermal designer must also be concerned with the fraction of heat generated within the fuel elements.

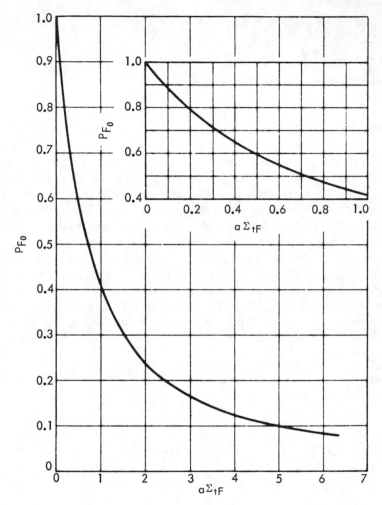

Fig. 1-24.   The probability, p_{F_0}, that a neutron escapes without collision from an infinite cylinder of radius a and macroscopic cross section Σ_{tF}. [From *J. Nucl. Energy*, **6**, 79 (1957).]

Fission energy is released in the form of kinetic energy of fission fragments, neutron kinetic energy, and gamma and beta rays. The very short range of fission fragments and beta particles ensures that this heat release will take place within the fuel elements.[c] The energy released during neutron thermalization is released to the moderator. The energy released on neutron capture is largely released within the fuel, but a portion of it is released within the reactor structure. The

---

[c] Approximately 10% of the beta energy is converted to gamma energy as bremsstrahlung and should be treated as gamma energy in the division.

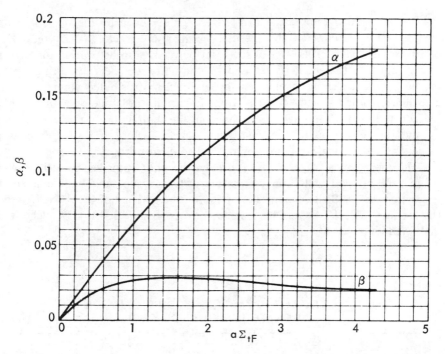

Fig. 1-25.   Parameters α and β as a function of $a\Sigma_{tF}$. [From *J. Nucl. Energy*, 6, 79 (1957).]

large mass of fuel produces a substantial self shielding, thus most gammas are captured within the fuel. Table 1-II quantitatively describes the distribution.

Five MeV of neutron kinetic energy are absorbed during thermalization. Hence, ~2.5% of the fission energy is definitely absorbed by the moderator. This can be taken advantage of in computing the heat that must be transferred by the fuel.

TABLE 1-II

ENERGY BREAKDOWN (MeV/fission) IN A WATER-COOLED REACTOR

Receiver	Time		
	Prompt	Delayed	Total
Fuel Element	Fission Fragment KE 168	β 7	175
Dispersed among fuel, moderator, and structure	Neutron KE 5 γ 7.5	Neutron capture 4 γ 6	22.5
Total	180.5	17	197.5

The portion of gamma and neutron energy absorbed by the moderator and core structure depends on the specific design of the reactor. In a typical PWR design, somewhat less than 1% of the total power is absorbed within the thermal shield; most of that energy is released by the outer fuel elements. In the central region of the core, the gamma energy is roughly distributed among the core constituents proportionally to their mass. In a uniform lattice core, this leads to the conclusion that essentially all gammas in the central core region are captured by the fuel elements. In a pressure tube reactor, which has a larger mass of moderator and internal structure, appropriate adjustment should be made.

The energy released per fission somewhat depends on fissile material as well as on control material and reactor structure. Walker[37] measured the total energy release for various fissile materials in pressure tube reactors of the CANDU design. His results (Table 1-III) indicate a value for ^{235}U fission that is ~2% higher than that listed in Table 1-II.

TABLE 1-III

**TOTAL ENERGY* RELEASED PER FISSION
IN REACTORS OF CANDU DESIGN**

^{232}Th	196.2 ± 1.1 MeV
^{233}U	199.7 ± 1.1 MeV
^{235}U	201.7 ± 0.6 MeV
^{238}U	208.5 ± 1.1 MeV
^{239}Pu	210.7 ± 1.2 MeV
^{241}Pu	213.8 ± 1.0 MeV

*Includes decay of absorption products but not energy of neutrinos.

1-3 POWER GENERATION AND DISTRIBUTION IN PRESSURE TUBE-TYPE CORES

1-3.1 Overall Power Distribution

In contrast to light-water-moderated reactors, the fuel in heavy-water or graphite-moderated reactors is generally not uniformly distributed throughout the moderator. As described in Sec. 1-1.1, the fuel elements in most designs are grouped together in clusters within pressure-containing tubes dispersed throughout the moderator. Nevertheless, an approximation of the overall distribution of thermal neutron flux, which is adequate for thermal design purposes, can be obtained by considering the moderator and fuel as homogeneous. However, we cannot normally consider these reactors to be bare since they are usually surrounded by effective radial reflectors. The effects of such a reflector can be approximately evaluated by considering two neutron energy groups—thermal and fast.

For a cylinder of infinite length, thermal flux within the core of a uniformly enriched reactor is given by[38]

$$\phi_t = AJ_0(\mu'r) + CI_0(\gamma r) \ ,$$

where A and C are constants, and μ' and γ are properties of the reactor core materials (see Ref. 34 for further discussion). For thin reflectors, the first term is dominant and flux distribution differs only slightly from that of an unreflected core. The change in flux pattern produced by a reflector around an infinite cylinder is illustrated in Fig. 1-26.

A somewhat cruder approximation of the behavior of a reflected core can be obtained by considering all neutrons to be thermal. When this is done, the resulting flux distribution equations have the same form as those for a bare core, but the overall dimensions are replaced by effective dimensions. For example, the flux of an unperturbed cylindrical core is approximated by

$$\phi = \phi_{max} J_0(2.405r/R_e)\left(\cos \frac{\pi z}{L_e}\right) , \qquad (1.23)$$

where

R_e = effective radius of core allowing for radial reflector

L_e = effective length of core allowing for axial reflectors.

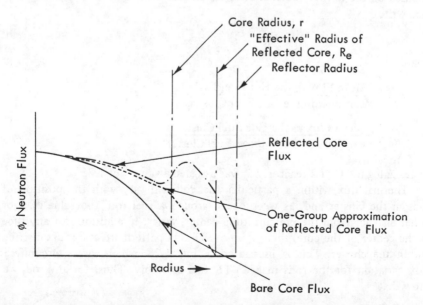

Fig. 1-26. Thermal neutron flux in bare and reflected cylindrical reactors.

Equation (1.23) does not predict the slight rise in core flux that occurs adjacent to the reflector. However, it can provide (Fig. 1-26) a reasonable estimate of flux throughout most of the core.

In the common cylindrical core, the large amount of structural material at the ends limits the effectiveness of any axial reflector. A cosine distribution with $L_e \approx L$ is thus often a very good approximation of the axial power variation in a uniformly loaded, unperturbed core similar to the CVTR.

1-3.2 Power Generation Within Fuel Clusters

The distance between rods within the pressure tube is generally orders of magnitude smaller than that between pressure tubes, resulting in negligible moderation within the pressure tube in comparison to that obtained between the tubes. As a first approximation, we therefore can ignore the area within the pressure tube as a source of thermal neutrons and approximate the thermal diffusion equation for thermal neutrons within the pressure tube by a modification of Eq. (1.13). Here we have

$$D_{av}\nabla^2\phi - \Sigma a_{av}\phi = 0 \ , \tag{1.24}$$

where D_{av} and Σa_{av}, respectively, represent the diffusion coefficient and macroscopic absorption coefficient based on homogenizing material within the pressure tube. From the development of Sec. 1-2.3, we see that this leads to an $I_0(R)$ flux distribution across the fuel cluster. Figure 1-27 illustrates this behavior in fuel clusters of the heavy-water-moderated and -cooled CVTR (Ref. 39).

Flux depression within the fuel assembly, called the "fine flux dip," is superimposed on the normal radial flux variation across the reactor as a whole. We can then write

$$F_R^N = F_G^N \times F_{R_0}^N \ , \tag{1.25}$$

where

$$F_G^N = \frac{\text{Mean Flux in the Hot Fuel Rod}}{\text{Mean Flux in the Hot Fuel Cluster}}$$

$$F_{R_0}^N = \frac{\text{Mean Flux in the Hot Fuel Cluster}}{\text{Mean Flux in the Average Fuel Cluster}} \ .$$

In designing the CVTR reactor, F_G^N was assigned a value[39] of 1.25.

Thermal flux within a particular fuel rod will vary with the position of the rods in the cluster and, as we proceed around a fuel rod, the value of R or the radial distance from the pressure tube center, varies. In addition, for any tube not in the center of the core, there is an overall flux pattern effect on flux distribution within the cluster. This is illustrated in Fig. 1-28, which shows circumferential flux variation for the rods in a CVTR fuel assembly. Thus, for any rod, we can show that

$$\phi = f(r,\theta) \ . \tag{1.26}$$

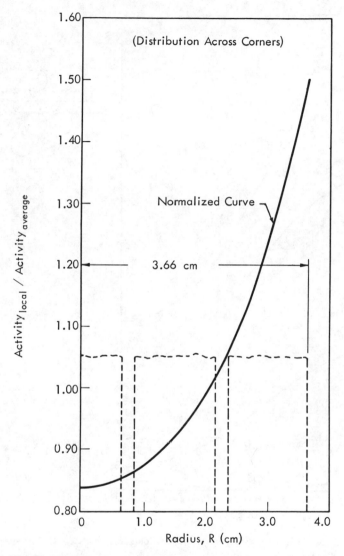

Fig. 1-27.   Flux distribution across radial axis of CVTR assembly (from Ref. 39).

To simplify temperature distribution calculation, distribution for a given rod is often approximated by a function of the form

$$\phi = C_1 + C_2 r \cos \theta \ . \tag{1.27}$$

The approximate function is made to fit actual distribution at the center of the rod and at the point on the circumference with the peak heat flux.[37] In a large core (CANDU designs), variation of the overall flux across a single pressure tube is generally small and angular power variation is of lesser importance.

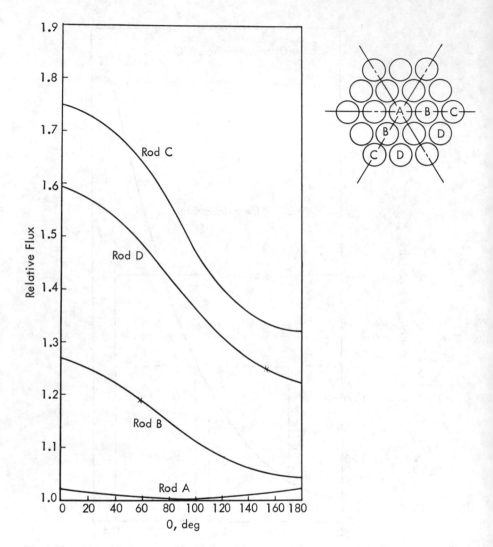

Fig. 1-28. Circumferential variation in flux around fuel rods of CVTR assembly (from Ref. 39).

1-3.3 Effect of Refueling Procedures

As previously noted, the CANDU reactor is the only PWR pressure tube reactor type of commercial interest. In this design, the large cylindrical calandria is usually placed with its axis horizontal and is pierced by a large number of once-through horizontal pressure tubes. Since these reactors are fueled with natural uranium, maximum burnups are low (~10 000 MWd/1000 kgU) and frequent refueling is necessary. Excessive downtime can be avoided by using on-power refueling.

Short fuel assemblies are used so that each channel contains a number of assemblies. When it is desired to add fuel to a channel, specially designed refueling machines are connected to each end of the channel. Fresh fuel assemblies are added at one end of the channel, while the refueling machine removes burned assemblies at the other end. By always adding fuel at the same end of a given channel, the fuel assemblies gradually move across the reactor. Thus, there is a significant burnup gradient along the axis of a given channel and, hence, axial power generation will not follow a simple cosine distribution. To prevent a power tilt in one direction, half the pressure tubes are loaded from one end of the reactor, while the others are loaded from the opposite end.

At BOL, a fuel assembly will be at the edge of the core and generate a low power. Later in life, it will see the high flux in the center of the reactor and operate at its maximum power. At the end-of-life (EOL), it again will be near the edge of the core and at low power. A typical fuel assembly operating history is shown in Fig. 1-29.

A realistic estimate of axial power distribution requires that variation in fuel burnup be considered. Since the burnup distribution is not axially continuous, axial power in a given channel will show a stepwise variation.

Since fresh fuel is added to each channel, the only refueling variable under the operator's control is the refueling rate, which determines discharge burnup from a given channel. In present (1978) fuel management practice, average discharge burnup is adjusted to give a flat radial flux. This leads to a nearly constant discharge burnup over the central region. The size of the inner region is adjusted to keep the maximum flux-to-average flux ratio within a prescribed limit. In the outer region, refueling rates are variable and are set to provide a discharge burnup (constant over the outer region) such that the reactor remains critical.

1-4 METHODS FOR RAPID ESTIMATION OF CORE POWER DISTRIBUTION

Accurate computation of core power distribution is usually accomplished by few-group diffusion theory calculations. The core is divided into a number of

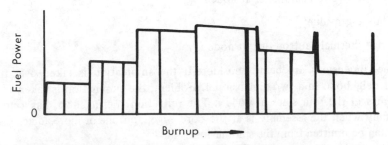

Fig. 1-29. Typical operating history of CANDU fuel assembly. [From *J. Nucl. Appl. Technol.*, 9, 195 (1970).]

small spatial regions in which number densities are assumed constant. Neutron balance equations are written for each neutron energy group at each spatial region. The equations are generally solved numerically by a complex inner and outer iteration scheme.[40,41] Local peaking within a homogenized region can be determined by a subsequent few-group calculation in which heterogeneity of the initially homogenized region is taken into account. In such a calculation, which is limited to a single small region with zero or near-zero current at the boundaries, fuel rods and moderator are explicitly represented by separate mesh points. Local peaking factors so determined are then superimposed on the previously determined power distribution.

In some thermal design procedures, it is desired to link thermal and nuclear calculations. Although it is possible to link thermal calculations directly with detailed reactor physics calculations, as in the THUNDER code,[42] this procedure is far too time consuming for general use. Less accurate, but more rapid, procedures for power distribution are often used under these circumstances.

It is common to consider radial core power distribution on an assembly-by-assembly basis. Power distribution estimations can be made on a two-dimensional basis with an assumed constant axial flux shape, or the assemblies can be divided into axial zones and a three-dimensional computation can be conducted. Radial peaking factors within a given assembly are usually based on values obtained in an initial few-group diffusion calculation.

The "nodal-coupling" method has been widely used for power distribution estimates. In FLARE (Ref. 43), the oldest nodal code, the neutron conservation equation, is written for each node as

$$\phi_i = (k_{\infty i}/\lambda) \left[\sum_{m=1}^{j}{}' W_{mi}\phi_m + \left(1 - \sum_{m=1}^{j}{}' W_{im}\right)\phi_i + \alpha_i \phi_i \right] , \qquad (1.28)$$

where

W_{mi} = probability that neutron born at node m is absorbed at node i

$k_{\infty i}$ = infinite lattice multiplication factor at spatial node i

α_i = reflection parameter at node i

λ = eigenvalue

ϕ_i = thermal neutron flux at node i.

If three dimensions are being considered, the summation is taken over the six nearest neighbors. In a two-dimensional problem, the summation is taken over the assemblies at the four sides of node i. The reflection coefficient α_i has a non-zero value only when the assembly is at the core edge, and one or more adjacent nodes must then be omitted from the summation.

In revised versions of FLARE (Ref. 44), the probability, W_{mi}, which is composed of transport and diffusion parameters, is given by

$$W_{mi} = \frac{(1 - g')\,(M_m^2)^{1/2}}{2(\Delta l)_{mi}} + \frac{g'\,(M_m)^2}{k_\infty (\Delta l_{mi})^2} \quad , \qquad (1.29)$$

where

M_m^2 = migration area

Δl_{mi} = node interval

g' = mixing factor (input parameter).

Accurate results from a nodal code require that input parameters g' (different g' can be used for horizontal and vertical directions) and α be fitted so initial agreement is obtained between the nodal code and few-group calculations. If an appropriate regression formula for the effect of voids, burnup, etc. on k_∞ is obtained, the nodal code can then be used to evaluate the effect of thermal conditions on flux and power. Somewhat better agreement can be obtained by considering coupling with diagonally adjacent nodes as done in the TRILUX code.[45]

Although both FLARE and TRILUX were originally devised for boiling water reactors where void effects are of major importance, nodal codes have been adapted for PWR use. Chitkara and Weisman[23] used a similar approach for the rapid power estimation needed in fuel shuffling studies. These authors considered g' to be unity in Eq. (1.29). They also included the effect of diagonal assemblies by extending the summation of Eq. (1.28) to include these assemblies. In evaluating W_{mi} for diagonally placed assemblies, Eq. (1.29) was multiplied by an obliqueness parameter to which they assigned a value of 0.7.

Chitkara and Weisman developed simplified relationships for P', the thermal power

$$P' = \phi_{th}[\sigma_5^{th} N_5 E_5 + \sigma_9^{th}(N_9 E_9 + N_1 E_1)]$$
$$+ r\phi_{th}[\sigma_5^f N_5 E_5 + \sigma_9^f(N_9 E_9 + N_1 E_1) + \sigma_8^f N_8 E_8] \quad , \qquad (1.30)$$

where

ϕ_{th} = thermal flux at given node

r = ratio of fast to thermal flux, assumed constant for purposes of power estimation

$\sigma_5^{th}, \sigma_9^{th}$ = thermal fission cross sections of ^{235}U and ^{239}Pu, respectively

$\sigma_5^f, \sigma_9^f, \sigma_8^f$ = fast fission cross sections of ^{235}U, ^{239}Pu, and ^{238}U, respectively

N_1, N_5, N_8, N_9 = number densities of ^{241}Pu, ^{235}U, ^{238}U, and ^{239}Pu, respectively

E_1, E_5, E_8, E_9 = fission energies liberated by ^{241}Pu, ^{235}U, ^{238}U, and ^{239}Pu, respectively.

For simplicity, we assume that $E_8 = E_5$ and $E_9 = 1.04\,E_5$. The number densities of the various fissile isotopes are described in terms of N, the number density of ^{238}U and b, the degree of burnup. The degree of burnup is defined by

$$db = \phi_{\text{th}}\sigma_{as}^{\text{th}}\,dt \;, \tag{1.31}$$

where t equals time and σ_{as}^{th} equals the thermal absorption cross section of ^{235}U.

At the end of a cycle, b_{fb}, the final degree of burnup is

$$b_{fb} = b_{in} + \sigma_{as}^{\text{th}}\int_0^T \phi_{\text{th}}dt \;, \tag{1.32}$$

where

$\quad T$ = cycle length

$\quad b_{in}$ = degree of burnup at BOC.

Chitkara and Weisman[23] concluded that b_{fb} could be approximated by

$$b_{fb} = b_{in} + T\sigma_{as}^{\text{th}}\bar{\phi} \;, \tag{1.33}$$

where

$\quad \bar{\phi}$ = effective average thermal flux $\simeq \phi_{in}\left(1 + \dfrac{1 + \phi_{in}}{3.0}\right)\phi_{\text{avg}}$

$\quad \phi_{in}$ = thermal flux at BOC

$\quad \phi_{\text{avg}}$ = average thermal flux across entire core, or radial slice if only a two-dimensional calculation.

The number density of ^{235}U is then obtained from

$$N_5/N_8 = (N_5/N_8)^{\circ}\,\exp(-b) \;, \tag{1.34}$$

where $(N_5/N_8)^{\circ}$ is the initial concentration of ^{235}U.

The fissile plutonium concentration is given by

$$N_{9,1}/N_8 = C_R\{(N_5/N_8)^8\,[1 - \exp(-b)]\,(1 - 0.25\,b)\} \;, \tag{1.35}$$

where C_R is the conversion parameter, which was found to be 0.43 for the San Onofre core.

To obtain accurate flux distributions with a nodal code, the user must carefully fit mixing parameters and reflection coefficients to achieve initial agreement with a multigroup calculation for some base loading of a core with the same configuration and power level. Some investigators[44] have used formal optimization procedures for this parameter determination. The tedium of this nodal approach can be avoided by using the 1.5-group coarse-mesh method.[24,46] Although much of the initial impetus for this approach came from BWR design,[46] the approach applies equally well to the PWR (Ref. 24).

In the 1.5-group approach, the fast-group diffusion theory equation is solved using finite difference equations. Thermal flux is then calculated assuming zero buckling in the thermal group. We assume that thermal group difference equations are not applicable since the 16- to 24-cm nodes are several thermal mean-free-paths apart. Therefore, thermal flux is calculated as if the mesh size were infinite for each node, and the thermal leakage from each node is taken as zero. In a two-dimensional calculation where only four adjacent cells are considered, fast flux ϕ_0^f at cell 0 is obtained from

$$\phi_0^f = \frac{S_0 V_0 + \sum_{k=1}^{4} \frac{\bar{D}_k A_k}{l_k} \phi_k^b}{\Sigma_0^R V_0 + \sum_{k=1}^{4} \frac{\bar{D}_k A_k}{l_k}} , \qquad (1.36)$$

where

S_0 = neutron source in cell 0

V_0 = volume associated with cell 0

l_k = distance between mesh point k and mesh point 0

A_k = area of the boundary separating mesh point k and mesh point 0

ϕ_k^f = fast flux at mesh point k

Σ_0^R = fast group removal cross section at 0.

Here, $\bar{D}_k$ is an effective diffusion coefficient defined by

$$\bar{D}_k = \frac{D_0 D_k (\delta_0 + \delta_k)}{D_0 \delta_k + D_k \delta_0} , \qquad (1.37)$$

where

D_i = diffusion coefficient for fast group at i

δ_i = width of cell i.

Fast flux is obtained from Eq. (1.36) by using standard inner and outer iteration methods with fission source updates after each iteration. Fission source S_0 is calculated by

$$S_0 = (1/k_{\text{eff}}) (\nu \Sigma_{fo} \phi_0 + \nu \Sigma_{\text{th}} \phi_0^{\text{th}}) , \qquad (1.38)$$

where

$\nu \Sigma_{fo}, \nu \Sigma_{f\text{th}}$ = fast and thermal fission source cross sections, respectively

ϕ_0^{th} = thermal flux at cell 0.

Thermal flux is obtained from

$$\Sigma_{a\text{th}} \phi_0^{\text{th}} = \Sigma_{so} \phi_0^f , \qquad (1.39)$$

where Σ_{s0} is the transfer scattering cross section from fast to thermal group at location 0.

Because of large mesh size spaces, the fast and thermal fluxes calculated from Eqs. (1.36) and (1.39) must be corrected to account for increases or decreases near the edge of the assembly. The average fast flux $\bar{\phi}$ is expressed as

$$\bar{\phi} = A'\phi_0^f + \frac{1-A'}{4} \sum_{j=1}^{4} \phi_j^i \ , \qquad (1.40)$$

where

ϕ_j^i = flux at the interface between nodes i and j

A^i = relative weight factor on the point flux of node i.

Borreson[46] found A' to be in the range of 0.7 to 0.76.

In the 1.5-group approach, the outside boundary condition is treated by including water nodes at the core boundary. The flux is set equal to zero at an appropriate distance into the water. The determination of reflection coefficients is thus avoided.

Stout and Robinson[24] compared 1.5-group calculations of the assembly average power distribution to that obtained from a multigroup fine-mesh calculation. They found a maximum error slightly in excess of 6%. In this calculation, all number densities were known.

1-5 TRANSIENT POWER GENERATION AND DISTRIBUTION

1-5.1 Power Distribution Following a Load Change

As previously noted, in a chemical-shim-controlled core, the control rods are nearly fully withdrawn at full power. However, if the plant is to be load following, rods would normally be inserted during part-load operation. Improper positioning of these rods for part-load operation can lead to an unacceptably high axial power peaking on return to full power.

When control rods are partially inserted, there is spatial change in fission product distribution. The distribution of ^{135}Xe is of particular importance because of its high yield and cross section (3×10^6 b). Randall and St. John[47] recognized this problem and pointed out that changes in xenon distribution following load changes can result in significant power shifts. More recent studies confirm this viewpoint.[33]

If all the reactivity change required for low-power operation is compensated for by rod motion, a deep insertion of the rods is required. This causes the flux to peak in the bottom half of the core and leads to reduced concentrations of ^{135}Xe and its precursor, ^{135}I, in the top half of the core. On return to full power by rod withdrawal, the initial ^{135}Xe concentration would largely be determined by the concentration of ^{135}I set at low-power operation. The power now peaks in the

upper portion of the core because of the lower ^{135}Xe concentration there. This
behavior, illustrated in Fig. 1-30, results in a higher axial peaking factor than
obtained under steady-state conditions.

The concept of "constant axial offset control" has been used to avoid increased
axial peaking due to xenon transients.[33] The axial offset, AO, can be defined as

$$AO = \frac{P'_T - P'_B}{P'_T + P'_B} \quad , \tag{1.41}$$

where

P'_T = total power generated in the upper half of the core

P'_B = total power generated in the bottom half of the core.

When a core operates at full power with all rods withdrawn, there is generally
an axial offset of somewhere between 0.0 and −0.1 (between 0.0 and −0.05 on all
but the first cycle). Axial offset varies slowly with lifetime. The equilibrium value
with rods withdrawn at any given point in life is called the "target" axial offset. It
has been found that if AO were required to remain at the target value through all
load changes, then on return to full power the axial peaking factor is not significantly changed.

If all the changes in reactivity occasioned by a load reduction were compensated for by boron addition, there would be an upward shift in core power; i.e.,

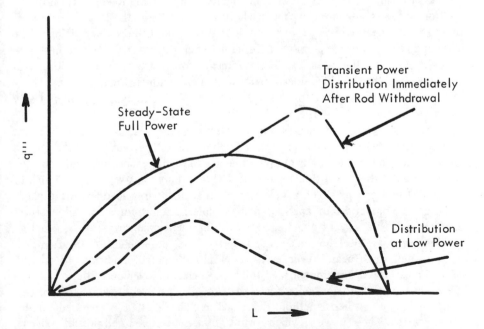

Fig. 1-30. Axial power shifts after control rod motion.

AO would become positive. This shift arises because of the local negative modera-tor coefficient effect and the larger temperature change near the top of the core compared to that in the bottom portion. Control rods are inserted at a level sufficient to prevent this power shift, and any additional reactivity reduction is achieved by the addition of boric acid. By using a combination of rod motion and boric acid concentration changes, load changes on the order of 1%/min can be accommodated. This rate of power change is sufficient to handle the daily load cycle to which a plant may need to respond.

Axial offset can be readily monitored by the reactor operator. In one scheme, ionization chambers that run the full length of the core are placed at four locations around the core circumference. Each chamber is split into an upper and lower portion and separate signals obtained from each section. Electronic comparison of these signals allows the axial offset to be read directly by the operator.

The concept of constant axial offset, which minimizes changes in power shape during core lifetime, is similar in some respects to the Haling principle.[48] This principle has been widely used for setting the optimum axial flux shape in BWRs. Haling states that rods compensating for reactivity changes with lifetime should be programmed to keep the axial power shape invariant over the in-core life of the fuel.

Limitations on the rate at which boric acid can be added and diluted do not allow a constant axial offset to be maintained using full length rods if a rapid (e.g., 5%/min) load change must be accommodated. Some plants have been equipped with "part-length" control rods in an effort to achieve rapid load follow-ing. A part-length rod contains poison material only in the lower portion of the control elements and an inert material, such as Al_2O_3, in the upper portion. If these rods are deeply inserted, they can depress the flux in the lower portion of the core and, thus, produce a more acceptable axial offset. Although it has been possible to meet the peak axial requirements for steady-state operation, such operation can lead to high local stresses on fuel cladding. Such stresses can be calculated as being in excess of those stress levels capable of causing clad failure by stress corrosion cracking in the presence of iodine.

Load following complicates plant safeguards studies. A reliable accident analysis begins with power distribution in the core at the onset of the accident. Since an accident could occur at any time and since the exact power distribution is a function of the power-time history of the core, there are a very large number of power distributions possible at the onset of an accident in a load-following plant. It becomes impractical to analyze each possible accident for all possible power distributions. One way of avoiding repetitive accident analyses is to construct an envelope of peak local power density (kW/ft) versus core height.[33] An attempt is made to determine peak power levels under all anticipated conditions by analyti-cally simulating all anticipated load-following behavior over core lifetime. From these many power distribution studies, the worst peak power at each axial position is selected. The synthesis method, described in Sec. 1-2.4, is often used for this determination. These powers, which form an upper envelope of all possible powers, are then used to describe core condition at the beginning of the accident. When accident analysis shows that the core design is safe for these conditions, then the core is clearly safe for less severe conditions that would actually exist. This approach is particularly useful in loss-of-coolant accident (LOCA) analyses.

1-5.2 Power Generation During Shutdown

1-5.2.1 Fission Power Generation

The rate of power decrease during a reactor shutdown is of concern to the thermal-hydraulic designer, since it affects fuel temperature during accident situations. We consider the situation where the reactor is shut down after operating at steady state for an appreciable period. During steady-state operation, the effective multiplication factor, k_{eff}, must be unity. For startup, it must be greater than unity. For shutdown, it must be less than unity. It is useful to define an excess multiplication factor k_{ex}, where

$$k_{ex} = k_{eff} - 1 . \tag{1.42}$$

Thus, k_{ex} is positive for a supercritical reactor and negative for a subcritical reactor. If all neutrons can be considered emitted promptly at the time of fission, the time behavior of the bare thermal reactor is given by

$$\phi(\bar{r},t) = \phi_0 \exp(k_{ex}\, t/l) , \tag{1.43}$$

where

ϕ_0 = steady-state neutron flux at $\bar{r}$

t = time since the power was suddenly changed by k_{ex}

l = mean lifetime of thermal neutrons.

Equation (1.43) can be considered valid immediately after the change. For longer periods, the effects of delayed neutron groups must be considered. If the delayed neutrons are approximated by a single group with a decay constant λ equal to the weighted average for the five groups, for small reactivity changes we obtain

$$\phi = \phi_0 \left\{ \frac{\beta}{\beta - \rho'} \exp\left(\frac{\lambda \rho' t}{\beta - \rho'}\right) - \frac{\rho'}{\beta - \rho'} \exp\left[-\frac{(\beta - \rho')t}{l} \right] \right\} , \tag{1.44}$$

where β is the total fraction of delayed neutrons and ρ', the reactivity, is defined as

$$\rho' = \frac{k_{ex}}{k_{eff}} . \tag{1.45}$$

The equation is valid only when $(\beta - \rho')$ is positive. Since l is on the order of magnitude of 10^{-3} s and λ is ~0.08 s^{-1}, the flux during shutdown is the sum of two positive terms (ρ' negative), one of which decreases much more rapidly than the other. The flux and, hence, power initially decreases rapidly and then falls off slowly in accordance with the first term within the bracket; subsequent slow decay is due to the effect of delayed neutrons.

The usual reactor shutdown is brought about by large reductions in reactivity. For this condition and considering all five groups of delayed neutrons, the neutron flux is closely approximated by

$$\phi = \phi_0 [A_1 \exp(-t\lambda_1) + A_2 \exp(-t\lambda_2) + \text{------} + A_5 \exp(-t\lambda_5) + A_6 \exp(-t/l)] \quad . \quad (1.46)$$

Here, λ are the decay constants for the five groups of delayed neutrons. For large reactivity decreases, the rate of flux decay is independent of the reactivity change. Noting the very small value of L and examining the values of λ in Table 1-IV (Ref. 49), we see that initially there is a rapid decrease in flux due to the last term, but after a short time, all terms become negligible except those with the smallest value of λ. Flux then decays exponentially with a period corresponding to the minimum $1/\lambda$, or ~ 80 s.

Based on the foregoing, a good approximation of the fission energy power following a large reactivity decrease is a sudden decrease (prompt jump) followed by a relatively slow exponential decay. The residual power of a light-water-moderated reactor shutdown, following sustained operation at constant power, can be approximated over the time period during which decay heat is important by

$$P'/P_0' = 0.15 \exp(-0.1\ t) \quad , \quad (1.47)$$

where t is in seconds.

Since the neutron lifetime on which the decay constant depends is largely determined by the reactor moderator, Eq. (1.47) should only be used for light water reactors. For heavy-water-moderated reactors, a decay constant of 0.06 should replace the 0.1. Also note that Eq. (1.47) does not apply to a plutonium-fueled core. The delayed neutron fraction for ^{239}Pu is only $\sim 0.21\%$ and, hence, the residual power will be $\sim \frac{1}{3}$ that of ^{235}U fuel.

1-5.2.2 Decay Heat

In addition to residual fission energy, there are two other major heat sources: (a) fission product radioactive decay, and (b) neutron capture product radioactive decay. Each of these sources must be considered.

TABLE 1-IV
DELAYED NEUTRONS IN THERMAL FISSION OF ^{235}U

Mean Life	Decay Constant [λ (s^{-1})]	Fraction (β_c)
0.62	1.61	0.00084
2.19	0.456	0.0024
6.50	0.151	0.0021
31.7	0.0315	0.0017
80.2	0.0124	0.00026

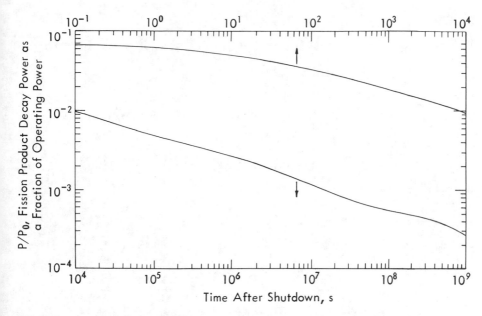

Fig. 1-31. Standard fission product decay heat curve for uranium-fueled thermal reactors (infinite operating time assumed).

Perkins and King,[50] Smith,[51] and Shure,[52] among others, considered fission product decay heat generation. The current (1978) design standard[53] is based on Shure's work.[52] Decay heat generation can be determined from the curves of Fig. 1-31 for reactors with an infinitely long operating time. The ordinate of this figure, P'/P_0', is the ratio of decay power to operating power. An upper design limit is obtained from these curves since they were obtained by adding an additional 20% to Shure's values to account for uncertainties. Schenter and England[54] compared Shure's curve to more recent decay heat computations and experiments. They conclude that the Shure +20% curve is conservative.

Schenter and Schmittroth[55] point out that Shure's decay heat evaluation is based on ^{235}U fission and decay heat depends on the fissionable isotope used. They observe that decay heat from ^{241}Pu fission products is very close to that for ^{235}U fission products for short cooling times and slightly below at long cooling times. Fission products of ^{239}Pu produce heats that are ~15% below the ^{235}U curve at zero time and the same at ~1000 s. Decay heat resulting from fast fission of ^{238}U is ~20% above the ^{235}U at zero time and slightly below at 1000 s. However, ^{238}U fast fission is only a very small fraction of the total fissions. The 20% uncertainty included in the design curve would appear to be very conservative and encompass any effects due to fissioning isotopes other than ^{235}U. Future design standards are likely to incorporate less conservatism.

The design standards require that the total decay energy release include contributions from ^{239}U and ^{239}Np. These contributions are:

For ^{239}U,

$$P'_{29}/P'_0 = 2.28 \times 10^{-3} C(\sigma_F/\sigma_{Ff}) [1 - \exp(-4.91 \times 10^{-4} t_0)] \exp(-4.91 \times 10^{-4} t_s) \ . \tag{1.48}$$

For ^{239}Np

$$P'_{39}/P'_0 = 2.17 \times 10^{-3} C(\sigma_F/\sigma_{Ff}) \{7.0 \times 10^{-3} [1 - \exp(-4.91 \times 10^{-4} t_0)]$$
$$\times [\exp(-3.41 \times 10^{-6} t_s) - \exp(-4.91 \times 10^{-4} t_s)]$$
$$+ [1 - \exp(-3.41 \times 10^{-6} t_0)] \exp(-3.41 \times 10^{-6} t_s)\} \ , \tag{1.49}$$

where

P'_{29} = decay power from ^{239}U

P'_{39} = decay power from ^{239}Np

C = conversion ratio, atoms of ^{239}Pu produced per atom of fissile material consumed

σ_F = effective neutron absorption cross section of fissile material

σ_{Ff} = effective neutron fission cross section of fissile material.

The conversion ratio, C, can be obtained from reactor physics calculations or estimated from

$$C = \frac{\sigma_{28}}{r'\sigma_F} + \eta\epsilon (1 - p)p_f \ , \tag{1.50}$$

where

σ_{28} = effective neutron absorption cross section of ^{238}U

r' = atom ratio fissile nuclei to ^{238}U

η = effective fission neutron yield per neutron absorbed in fissile material

ϵ = fast fission factor

p = resonance escape probability

p_f = fast neutron nonleakage probability.

Decay heat due to capture in control material varies greatly depending on the nature of the absorber present in the core during power operation. Decay heat due to silver-indium-cadmium control rods would be more significant than that due to boron-stainless-steel rods. In any case, the capture product decay heat is substantially less than that due to capture in ^{238}U, but no general expression for this heat generation is available.

The contribution from capture in the structure is small. It is generally satisfactory to ignore this.

If the residual heat generation after several days of shutdown were of interest, the fact that the reactor has operated for a finite time rather than an infinite time becomes important. Obenshain and Foderaro[56] state that residual heat for a finite time of operation (T_0) can be found by subtracting the value of $(t + T_0)$ from the value for infinite operation with a true decay time of t. That is,

$$\frac{P'}{P_0'}(t_0, t_s) = \frac{P'}{P_0'}(\infty, t_s) - \frac{P'}{P_0'}(\infty, t_0 + t_s) , \qquad (1.51)$$

where

$\dfrac{P'}{P_0'}(t_0, t_s)$ = decay heat ratio for a reactor that has operated for time t and been shut down for time t_s

$\dfrac{P'}{P_0'}(\infty, t_s)$ = decay heat ratio for a reactor operated for an infinite time and shut down for time t_s

$\dfrac{P'}{P_0'}(\infty, t_0 + t_s)$ = decay heat ratio for a reactor operated for an infinite time and shut down for time $(t_0 + t_s)$.

Also note that for decay times in excess of one minute, the contribution of residual fission energy generally can be ignored.

By combining the estimates from Fig. 1-31 with those of Eqs. (1.47), (1.48), and (1.49), we estimate the total power generation at time t after a reactor shutdown. Figure 1-32 shows the relative power generation for a typical light-water-moderated reactor, where the heat released to the coolant is appreciably greater

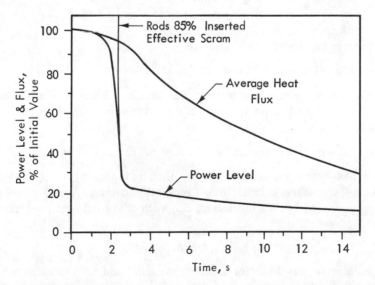

Fig. 1-32. Relative power level and heat flux after a scram. [From *Westinghouse Engineer*, **25**, *5*, 145 (1965).]

than the heat generated, due to the release of heat stored in the fuel. Means for taking this into account are discussed in Sec. 2-3.

1-6 HEAT GENERATION IN REACTOR STRUCTURE AND MODERATOR

1-6.1 Structure

Heat is produced in the fuel by absorption of beta rays and the kinetic energy of the fission fragments, while outside the fuel, heat is generated primarily by gamma-ray absorption. Gamma rays come from gammas released in fission, those released by decaying fission products, or those formed by absorption of neutrons in the core materials.

The variation of the uncollided flux of gamma rays ϕ^0 of a given energy through a slab with a source on one side is given by

$$\phi_i^0 = S_{a_i} \exp(-\mu_i x) , \qquad (1.52)$$

where

S_{a_i} = flux at the surface, MeV/s cm^2

x = distance from surface, cm

μ_i = attenuation coefficient, cm^{-1}.

The associated heat production is then

$$q_i^0 = \mu_{a_i} S_{a_i} \exp(-\mu_i x) , \qquad (1.53)$$

where

μ_{a_i} = energy absorption coefficient, cm^{-1}

q_i^0 = heat production, MeV/s cm^2.

In a reactor where we have a source emitting rays of a number of different energies, the total heat production due to uncollided flux is the sum of heat production due to each energy level.

$$q_t^0 = \mu_{a_1} S_{a_1} \exp(-\mu_1 x) + \mu_{a_2} S_{a_2} \exp(-\mu_2 x) + \ldots . \qquad (1.54)$$

The actual heat production is greater than indicated by Eq. (1.54) due to absorption of secondary gammas from Compton scattering. Additional heat production is computed by buildup factors B_{a_i} such that total heat production q is given by

$$q = q_1^0 B_{a_1} + q_2^0 B_{a_1} + q_2^0 B_{a_2} + \ldots . \qquad (1.55)$$

The buildup factors differ for different materials and are a function of μx and the energy level of the gamma rays. Rockwell[57] gives a graphic presentation of these.

In a reactor core that has been operating for 100 h or more, the gamma-ray energy from both fission and fission products is assumed to be directly proportional to the local power density. Values for gamma-ray energy production from fission and fission products, as well as energy attenuation coefficients, are summarized by McLain and Martens.[58] Where gamma flux leaving the core must be determined, power density near the surface of the core can be used as an approximation, since high self-shielding of the core minimizes contributions from other regions.

The variation of neutron flux through a thin slab with a source on one side is of the same form as that for gamma rays

$$\phi = \phi_0 \exp(-\Sigma x) \ , \tag{1.56}$$

where

ϕ = local neutron flux at x

ϕ_0 = neutron flux at surface of slab

Σ = neutron absorption coefficient, cm^{-1}.

In a structural material consisting of heavy elements, neutron heat production is primarily due to absorption of gammas produced by neutron interaction with the structure. Heat production q''', due to capture gamma rays of a given energy level, is therefore directly proportional to the neutron flux and a function of μ_a, x, and total slab thickness. If the neutron flux is given by Eq. (1.56), then

$$q'''_E = KEf(E)\mu_a(\Sigma/\mu)\phi_0 \exp(-\Sigma x)\{F_1(\mu x, \Sigma/\mu) + F_1[\mu(L-x), -\Sigma/\mu]\} \ , \tag{1.57}$$

where

q'''_E = heat production due to capture gamma-ray absorption, W/cm^3

$f(E)$ = fraction of captures producing gamma rays of energy E

L = total slab thickness

$K = 0.8 \times 10^{-13}$ Ws/MeV

μ = linear attenuation coefficient for gammas of energy E, cm^{-1}

μ_a = energy absorption coefficient for gammas of energy E, cm^{-1}

and

$$F_1(b,a) = \int_0^b \exp(at) E_1(t)dt \ ; \ E_1(b) = \int_b^\infty \frac{\exp(-t)}{t} dt \ , \tag{1.58}$$

where b and a are the appropriate arguments required by Eq. (1.57). Function F_1 can be evaluated from Rockwell's[59] graphs. Equation (1.57) neglects buildup of gamma rays due to Compton scattering; therefore, these results should be modified by application of the appropriate buildup factor for thick elements.

Total heat generation is then the sum of these values and those obtained from Eq. (1.55).

For steel slab thickness under 1.5 cm, it is usually satisfactory to assume a constant heat production due to neutron absorption. In this range, buildup factors approximately compensate for neutron flux attenuation. Furthermore, in the slabs with thicknesses between 1.5 and 6 cm, absorption capture gammas account for <10% of total heat production. This heat production can be approximated as a constant fraction of total heat production. Note that in a pressure vessel wall adjacent to a shield tank, the neutron flux begins to rise near the shield tank due to the reflected flux. Here, a better approximation would be constant heat production due to gamma absorption.

1-6.2 Moderator

While heat generation in the reactor structure is primarily from gamma-ray absorption, heat generation within the moderator is primarily from neutron thermalization through elastic scattering. For elastic scattering $q'''(x)$, the volumetric heat generation rate at some location x can be approximated by[60]

$$q'''(x) \approx \Sigma_s \phi_f(x) \xi \ , \qquad (1.59)$$

where

Σ_s = macroscopic elastic scattering cross section of the fast neutron flux $\phi_f(x)$

ξ = average energy decrement per collision.

1-7 THERMAL DESIGN BASIS

In view of the many variables that affect power generation within the reactor fuel, it is clear that an accurate computation of the relative power distribution in the reactor is quite complicated. The simple approximations presented in this chapter for uniform, unperturbed cores can be regarded as useful in providing very rough estimates for preliminary design. However, any actual design must be based on realistic power distributions obtained from the nuclear designer. The major results of the nuclear calculations can be relayed to the thermal designer in terms of nuclear hot-channel factors. These factors, with an estimate of the axial power-distribution shape, can serve as a basis for initial plant thermal design. Because of the wide use of chemical shim control, simple relationships are often adequate to describe axial power shape.

Refined analyses of plant performance may wish to consider thermal and nuclear effects simultaneously. This can be accomplished by linking the approximate methods for power estimation from Sec. 1-1.4 with the appropriate thermal design procedures.

REFERENCES

1. G. H. Farbman, "The Pressurized Water Reactor," Paper No. 65-15, Westinghouse Nuclear Power Seminar (1965).

2. J. K. Pickard, Ed., *Nuclear Power Reactors*, pp. 70-73, D. Van Nostrand Company, Princeton, New Jersey (1957).

3. H. J. von Hollen and C. F. Currey, "Design Features of San Onofre Nuclear Station," *Westinghouse Engineer*, **25**, *5*, 145 (1965).

4. E. U. Powell and R. M. Harper, "R. E. Gina Nuclear Power Plant Unit No. 1," *Proc. Am. Power Conf.*, **30**, 298 (1968).

5. Stuart McLaine and J. H. Martens, Eds., *Reactor Handbook*, 2nd ed., Vol. IV, pp. 557-560, Wiley-Interscience Publishers, New York (1964).

6. W. D. Fletcher and D. D. Malinowski, "Operating Experience with Westinghouse Steam Generators," *Nucl. Technol.*, **28**, 356 (1976).

7. D. K. Davies, "Nuclear Steam Generators," *Proc. Am. Power Conf.*, **27**, 310 (1965).

8. D. Kalafati, "Thermodynamic Cycles of Nuclear Power Stations," Israel Program for Scientific Translation (1965).

9. F. A. Artusa, "Turbines and Cycles for Nuclear Power Plant Application," *Proc. Am. Power Conf.*, **29**, 280 (1967).

10. "Consolidated Nuclear Steam Generator for Merchant Ship Application," BAW-1243, Babcock & Wilcox Company (Aug. 1962).

11. International Atomic Energy Agency, *Directory of Nuclear Reactors, Vol. IV, Power Reactors*, pp. 39-45, Vienna (1962).

12. International Atomic Energy Agency, *Directory of Nuclear Reactors, Vol. IV, Power Reactors*, pp. 163-168, Vienna (1962).

13. International Atomic Energy Agency, *Directory of Nuclear Reactors, Vol. IV, Power Reactors*, pp. 169-174, Vienna (1962).

14. International Atomic Energy Agency, *Directory of Nuclear Reactors, Vol. IV, Power Reactors*, pp. 157-162, Vienna (1962).

15. E. Critoph. S. Banerjee, F. W. Barclay, D. Hamel, M. S. Milgram, and J. I. Veeder, "Prospects for Self-Sufficient Equilibrium Thorium Cycles in CANDU Reactors," *Trans. Am. Nucl. Soc.*, **22**, 706 (1975).

16. International Atomic Energy Agency, *Directory of Nuclear Reactors, Vol. IV, Power Reactors*, pp. 3-8, Vienna (1962).

17. W. J. Dowis, "Basis of Design for Hanford New Production Reactor (NPR)," HW-SA-2981, Hanford Works (Apr. 1963).

18. J. A. Wright, "Supercritical Pressure Reactor Technology," Paper No. 65-8, Westinghouse Nuclear Power Seminar (1965).

19. S. Glasstone and M. C. Edlund, *Elements of Nuclear Reactor Theory*, p. 198, D. Van Nostrand Company, Princeton, New Jersey (1952).

20. S. Glasstone and M. C. Edlund, *Elements of Nuclear Reactor Theory*, p. 215, D. Van Nostrand Company, Princeton, New Jersey (1952).

21. "1000 MW(e) Closed Cycle Water Reactor Study," WCAP-2385, Vol. I, Westinghouse Atomic Power Division (1963).

22. B. Rothleder, "Power Reactor Fuel Management Beyond First Cycle Operation," *Trans. Am. Nucl. Soc.*, 14, 89 (1971).

23. K. Chitkara and J. Weisman, "An Equilibrium Approach to Optimal In-Core Fuel Management for Pressurized Water Reactors," *Nucl. Technol.*, 24, 33 (1974).

24. R. B. Stout and A. H. Robinson, "Determination of Optimum Fuel Loading in Pressurized Water Reactors Using Dynamic Programming," *Nucl. Technol.*, 20, 73 (1973).

25. B. N. Naft and A. Sesonske, "Pressurized Water Reactor Optimal Fuel Management," *Nucl. Technol.*, 14, 123 (1972).

26. International Atomic Energy Agency, *Directory of Nuclear Reactors, Vol. IV, Power Reactors*, pp. 21-32, Vienna (1962).

27. J. K. Pickard, Ed., *Nuclear Power Reactors*, pp. 77-80, D. Van Nostrand Company, Princeton, New Jersey (1957).

28. O. Schulze, "Matrix Solutions for Reflected Reactors," *Nuclear Engineering Handbook*, H. Etherington, Ed., pp. 6-70, McGraw-Hill Book Company, New York (1958).

29. A. Radkowsky and R. T. Bayard, "Physics of Seed and Blanket Cores," *Proc. 2nd U.N. Conf. Peaceful Uses At. Energy*, Vol. 13, p. 128, United Nations, New York (1958).

30. J. K. Pickard, Ed., *Nuclear Power Reactors*, p. 111, D. Van Nostrand Company, Princeton, New Jersey (1957).

31. "Shippingport LWBR Preliminary Safeguards Analysis Report," Bettis Atomic Power Laboratory (1975).

32. G. N. Hamilton and J. A. Roll, "Power Distribution and Reactivity Measurements in Critical Lattices Containing Thimble Control," WCAP-1894, Westinghouse Atomic Power Division (1964).

33. T. Morito, L. R. Scherpell, K. J. Dzikowski, R. E. Radcliffe, and D. N. Lucoff, "Topical Report on Power Distribution Control and Load Following Procedures," WCAP-8403, Westinghouse Atomic Power Division (1974).

34. S. Glasstone and M. C. Edlund, *Elements of Nuclear Reactor Theory*, p. 266, D. Van Nostrand Company, Princeton, New Jersey (1952).

35. A. Amouyal, P. Benoist, and J. Horowitz, "Nouvelle Méthode de Détermination du Facteur d'Utilization Thermique d'une Celle," *J. Nucl. Energy*, 6, 79 (1957).

36. J. R. Lamarsh, *Introduction to Nuclear Reactor Theory*, pp. 382-389, Addison-Wesley Publishing Company, Inc., Reading, Massachusetts (1966).

37. W. H. Walker, "Mass Balance Estimate of Energy Released per Fission in Reactors," AECL-3109, Atomic Energy of Canada Ltd. (1968).

38. S. Glasstone and M. C. Edlund, *Elements of Nuclear Reactor Theory*, p. 243, D. Van Nostrand Company, Princeton, New Jersey (1952).

39. A. A. Bishop, P. Nelson, and E. A. McCabe, Jr., "Thermal and Hydraulic Design of the CVTR Fuel Assembly," CVNA-115, Westinghouse Atomic Power Division (June 1962).

40. A. F. Henry, *Nuclear Reactor Analysis*, MIT Press, Cambridge, Massachusetts (1975).

41. M. K. Butler and J. M. Cook, "One-Dimensional Diffusion Theory," *Computing Methods in Reactor Physics*, H. Greenspan, C. N. Kelber, and D. Okrent, Eds., Gordon and Breach Science Publishers, Inc., New York (1968).

42. W. T. Sha, S. M. Hendley, and G. H. Minton, "An Integral Calculation of Three-Dimensional Nuclear-Thermal-Hydraulic Interaction," *Trans. Am. Nucl. Soc.*, 9, 476 (1966).

43. D. L. Delp, D. Fisher, J. M. Harrema, and M. J. Stedwell, "FLARE: A Three-Dimensional Boiling Water Reactor Simulator," GEAP-4598, General Electric Company (1964).

44. T. Kiguchi and T. Kawai, "A Method for Optimum Determination of Adjustable Parameters in the Boiling Water Reactor Core Simulator Using Operating Data on Flux Distribution," *Nucl. Technol.*, 27, 315 (1975).

45. L. Goldstein, F. Nakachi, and A. Veras, "Calculation of Fuel-Cycle Burnup and Power Distribution of Dresden I Reactor with the TRILUX Fuel Management Program," *Trans. Am. Nucl. Soc.*, 10, 300 (1967).

46. S. Borreson, "A Simplified Coarse-Mesh Three-Dimensional Diffusion Scheme for Calculating Gross Power Distribution in a Boiling Water Reactor," *Nucl. Sci. Eng.*, 44, 37 (1971).

47. D. Randall and S. St. John, "Xenon Spatial Oscillations," *Nucleonics*, 16, 3, 82 (1958).

48. R. K. Haling, "Operating Strategy for Maintaining Optimum Power Distribution Through Life," USAEC Report TID-7672, Division of Technical Information, Oak Ridge, Tennessee (1963).

49. S. Glasstone and M. C. Edlund, *Elements of Nuclear Reactor Theory*, p. 65, D. Van Nostrand Company, Princeton, New Jersey (1952).

50. J. F. Perkins and R. W. King, "Energy Release from Decay of Fission Products," *Nucl. Sci. Eng.*, 3, 726 (1958).

51. M. R. Smith, "The Activity of Fission Products of U-235," AEC Report XDC-60-1-157, General Electric Company (1959).

52. K. Shure, "Fission Product Decay Energy," WAPD-BT-24, Bettis Technical Review, Westinghouse Electric Company (1961).

53. American Nuclear Society Draft Standard 5.1, "Decay Energy Release Rates Following Shutdown of Uranium-Fueled Thermal Reactors," revised October 1973.

54. R. E. Schenter and T. R. England, "Nuclear Data for Calculation of Radioactivity Effects," *Trans. Am. Nucl. Soc.*, 22, 517 (1975).

55. R. E. Schenter and F. Schmittroth, "Radioactive Decay Heat Analysis," *Proc. Conf. Neutron Cross Sections and Technology*, Washington, D.C. (1975).

56. F. E. Obenshain and A. Foderaro, "Energy from Fission Product Decay," WAPD-P-652, Westinghouse Atomic Power Division (1955).

57. T. Rockwell, *Reactor Shielding Design Manual*, pp. 429-433, D. Van Nostrand Company, Princeton, New Jersey (1956).

58. Stuart McLain and J. H. Martens, Eds., *Reactor Handbook*, 2nd ed., Vol. IV, pp. 429-433, Wiley-Interscience Publishers, New York (1964).

59. T. Rockwell, *Reactor Shielding Design Manual*, pp. 353-358, D. Van Nostrand Company, Princeton, New Jersey (1956).

60. S. Glasstone and A. Sesonske, *Nuclear Power Engineering*, p. 615, D. Van Nostrand Company, Princeton, New Jersey (1967).

2

FUEL ELEMENTS

The thermal design of a reactor can be considered to begin with the design of the heat generating elements. As its first step, prediction of fuel element behavior in a reactor requires a knowledge of fuel temperatures during steady-state and transient conditions. Therefore, for the fuels of interest, we first review the properties necessary for temperature estimation and then examine some of the computational methods involved.

2-1 FUEL ELEMENT CHARACTERISTICS

2-1.1 Fuels and Their Thermal Properties

2-1.1.1 Metallic Alloy Fuels

Radiation damage to metallic fuel elements is caused largely by highly energetic fission fragments. The damage mechanisms fall into four categories:

1. displacement of individual atoms (the primary knock-on atoms) from the normal lattice

2. displacement of additional atoms by the primary knock-on atoms to form displacement spikes

3. the presence of fission products as impurities

4. cavity formation at points of mechanical weakness.

These changes result in both axial growth and radial swelling of the fuel.

Pure uranium metal is not a satisfactory fuel for PWRs designed for power production alone, since it exhibits very poor dimensional stability under prolonged irradiation and its high corrosion rate would result in rapid fuel element rupture following any cladding failure. Exposure of the failed fuel to 550°F water for a few hours would completely destroy the fuel element. On the other hand, when reactor operation is for both plutonium production and power, uranium metal is acceptable because its radiation stability is adequate during the short reactor exposure designed to limit higher isotope production. Coolant temperatures can be significantly reduced so that the consequences of any failures will be less severe.

Much of uranium's irradiation instability is traceable to the anisotropy of its alpha phase, which leads to expansion in the direction of the (010) crystal plane and contraction in the (100) plane during low-temperature irradiation. A substantial reduction in anisotropic behavior can be obtained by beta heat treating processes using oil quenching.[1] Further improvement in uranium properties can be obtained by adding a sufficient amount of an alloying element so the gamma phase can be retained at low temperatures. The gamma phase has a cubic structure that does not exhibit anisotropic properties.

Considerable effort was expended in attempts to develop alloys fully suitable for PWR service. The most promising were binary alloys of uranium and molybdenum.[2] Molybdenum is capable of delaying the beta-alpha transformation when present as a dilute (<5%) alloying addition. In moderate amounts (>5%), it can retain the metastable gamma phase at room temperature. Gamma-stabilized alloys were found to have significantly better aqueous corrosion resistance than pure uranium; alloys containing 10 to 13.5% molybdenum appear to exhibit adequate corrosion resistance for power reactor use when covered by cladding. Their dimensional stability is better than that of unalloyed uranium; however, while considerably improved, swelling remained a problem.[3] Interest was also shown in gamma-phase alloys of uranium, niobium, and zirconium. A uranium-molybdenum alloy fuel was initially considered for the blanket of the first Shippingport core, but the superior performance of uranium dioxide led to the use of ceramic fuel instead.

Uranium-molybdenum alloys have been successfully used as a fuel for pulsed test reactors.[4] The fuel for such reactors must be able to withstand intense shock loading and high dynamic stresses. Total burnup tends to be low; fuel swelling is not expected to prove a problem. The gamma-phase U–10 wt%–Mo alloy has been the most widely used alloy for this purpose.[4]

Attention has also been given to developing very low-alloy uranium fuels for a heavy-water-moderated PWR. The greater neutron economy achievable with low-alloy uranium is highly important in such reactor systems. Furthermore, the consequences of fuel element failure are less severe. Since heavy water units are pressure tube reactors, refueling is easier and severe element failures will not propagate throughout the reactor as they might in the close-packed lattice of a light water reactor.

The concept employed has been to form alloys using small amounts of suitable metals that tend to stabilize the gamma or beta phase. When transformation to the alpha phase takes place, the resulting material consists of small, randomly oriented crystals. The alloys have better dimensional stability and corrosion resistance than cast alpha-phase uranium. The alloys examined in depth have included U–2 to 2.5 wt%–Zr, U–0.3 wt%–Cr, U–1.5 wt%–Mo, and U-Fe alloys.[5,6] The so-called "adjusted uranium" alloy that contains ~500-ppm iron, 250-ppm aluminum, 600-ppm carbon, and 20-ppm nitrogen has also been studied.

An extensive study of the swelling of low alloys containing varying amounts of iron, silicon, aluminum, and molybdenum has been conducted during the development of an improved Savannah River reactor fuel.[1] Alloys containing 250- to 350-ppm silicon alone, or in conjunction with other alloying elements—especially molybdenum—were most effective for exposures up to 5000 MWd/tonne. Alloys containing 350-ppm iron, 350- and 800-ppm aluminum had satisfactory swelling behavior at temperatures below 400°C and exposures below 13 000 MWd/tonne.

While it has been possible to reduce fuel swelling substantially through alloy addition, the problem has not been completely eliminated. Below ~640°C, the swelling obtained varies greatly with alloy content; this effect is consistent with the concept that at these temperatures swelling occurs by mechanical processes. The benefits obtained from alloying additions appear to be due to the influence of the additions on mechanical properties. A minimum swelling rate is reached in the 640 to 700°C range. At temperatures above this, the rates again increase, but are only slightly different for the various alloys.[7]

The Canadians investigated the use of U_3Si (U—3.8 wt%—Si) for PWR service. While swelling is considerably lower than that of unalloyed uranium, it is higher than desired. However, the fuel is sufficiently plastic so the addition of a central void appears to offer a way of accommodating a major portion of the swelling.[8]

The water corrosion resistance of uranium is significantly improved by the addition of small amounts of molybdenum, zirconium, Nb + Zr, or silicon. Indeed, U_3Si has a corrosion resistance perhaps 500 times that of unalloyed uranium. However, attack of a fuel element of these alloys with a cladding hole leak would be severe enough to require fairly prompt removal of the element from the reactor. Small aluminum additions, e.g., 1.5% aluminum, substantially improve the corrosion resistance of U_3Si (Ref. 9).

A major inhibition to the use of uranium silicide is that U_3Si undergoes rapid and gross swelling at fuel temperatures in excess of 900°C (Ref. 9). Such temperatures can be encountered during a LOCA. Expected external pressures are not adequate to prevent severe ballooning. Mathews and Swanson[9] suggest that such high-temperature swelling may be typical of all metallic fuels with fission gases in irradiation-induced pores and may pose a severe limitation on the use of metallic fuels in reactors where accident temperatures approach the fuel melting point.

High-alloy uraniums with more satisfactory properties are available. Such alloys provide a means for dispersing highly enriched uranium in a structure with a sufficiently large surface area to attain the high heat removal rates required. The only high-alloy uraniums known to be used in PWRs are alloys of uranium and zirconium. Zirconium's low neutron absorption cross section, high melting point, and corrosion resistance, and uranium's high solubility in zirconium make it a natural alloying material. Zirconium alloys have adequate aqueous corrosion resistance and adequate strength when they contain a small amount of tin. In actual practice, uranium is alloyed with Zircaloy-2, which contains 1.5% tin plus traces of iron, nickel, and chromium. Dimensional stability under irradiation is good. Alloys containing up to 14 wt% U have satisfactorily withstood burnups of up to 3.3 at.%. Maintenance of dimensional stability seems to require that the fuel not be cycled above the phase transition temperature ($\approx$600°C) (Ref. 10). An alloy of uranium and Zircaloy was used for the meat of plate-type sandwiches in the first Shippingport reactor seeds.

The thermal design properties of U-Zr alloys were investigated by Battelle Memorial Institute. Etherington[11] tabulates the thermal conductivity of the alloy and pure components as a function of alloy composition and temperature. Density of the alloy at room temperature is approximately a linear function of the atomic percentages of the constituents.[12] Similarly, specific heats can be estimated from that of the components. Further properties can be obtained from Ref. 13. A summary of significant properties is given in Table 2-I.

The thermal properties of U-Mo alloys have also been extensively investigated. Room-temperature thermal conductivity varies from 0.148 W/cm °C for 8 wt% Mo to 0.138 W/cm °C for 12 wt% Mo (Ref. 13). The thermal conductivity of 8 wt% Mo fuel increases linearly with temperature to 0.39 W/cm °C at 790°C (Ref. 14). Thermal expansion is not linear with temperature, but shows an inflection at 595°C. A total length change of 1.8% is seen at 790°C (Ref. 14). Additional properties can be obtained from Ref. 14.

Properties of the ternary uranium, 7.5 wt% Mo, 2.5 wt% Nb, alloy are reviewed by Weiss.[15] Properties of a number of other alloys of interest are available in Refs. 13 and 16. Thermal properties for the most important alloys are summarized in Table 2-I.

Metallic thorium containing small amounts of enriched uranium has been proposed as a PWR fuel.[17] Thorium has a face-centered cubic (fcc) structure and does not exhibit the anisotropic behavior of uranium. The irradiation experiments available (1978) give no indication of instability and restrained swelling is 1 to 2% for 10 000 MWd/tonne exposure at 650°C (Ref. 17). The corrosion rate of pure thorium in 300°C water is two orders of magnitude lower than that of uranium. Addition of up to 6% uranium does not appear to affect corrosion rate. A further reduction in corrosion rate is achieved by small additions (0.03 to 0.12 wt%) of carbon. The carbon addition also leads to a more than three-fold increase in yield stress.

Thorium is a semirefractory metal (melting point of 1845°C) and has a good thermal conductivity ($\sim$0.1 cal/cm s °C). Hence, fuel temperatures during a LOCA should be far below the melting point and disastrous swelling would not be expected. However, no direct tests of this appear to be available (1978). Use of thorium metal instead of ThO_2 in an LWBR is predicted to lead to a slight improvement in breeding ratio and a reduction in seed enrichment.[17]

2-1.1.2 Dispersion Fuels

An alternative way to provide a structural base with the necessary high surface for highly enriched fuel is to physically disperse the fuel in a metallic matrix. Dispersions of UO_2 in aluminum, zirconium alloys, and stainless steels have been investigated, but only stainless-steel and zirconium alloy dispersions have been used in PWRs.

TABLE 2-I

THERMAL PROPERTIES OF URANIUM ALLOYS

Fuel	Uranium	U–2 wt% Zr	U_3Si (U–3.8 wt% Si)	U–12 wt% Mo	Zr–14 wt% U
Fuel density (g/cm^3)	19.05 (200°F) 18.87 (400°F) 18.33 (1200°F)	18.3 (room temperature)	15.57 ± 0.02 (room temperature)	16.9 (room temperature)	7.16 (room temperature)
Crystal structure	α, room temperature–1200°F, orth. β, 1200 to 1238°F, tet. γ, 1400°F-mp, body-centered cubic (bcc)	Orth.	$U + U_3Si_2$ > 1706°F ε < 1706°F tet. (ε formed by holding cast alloy at 1472°F)	bcc	α Zr (hex.) ε Zr (hex.)
Melting temperature (°F)	2070	2060	1805	2102	3240
Conductivity [Btu/(h ft °F)]	15.8 (200°F) 17.5 (600°F) 20.25 (1000°F) 22.00 (1400°F)	12.7 (95°F) 15.6 (572°F) 21.4 (1112°F) 27.8 (1652°F)	8.66 (77°F) 10.1 (149°F)	7.97 (room temperature)	6.36 (68°F) 6.71 (212°F) 7.12 (392°F) 7.52 (572°F) 10.4 (1292°F)
Thermal expansion [in./(in. °F)]	20.2×10^{-6} for (100) plane -5.17×10^{-6} for (010) plane 77°F 19.0×10^{-6} for (001) plane to 1202°F Volumetric 34.2×10^{-6}	8.0×10^{-6} (105 to 930°F)	$7.67 \pm 0.28 \times 10^{-6}$ (212 to 752°F)	7.23×10^{-6} (212 to 752°F)	3.78×10^{-6} (221 to 626°F) 3.84×10^{-6} (662 to 1022°F)
Heat capacity [Btu/(lb °F)]	0.0278 (200°F) 0.0410 (1000°F) 0.0464 (1200°F)	0.0287 (200°F)		0.032 to 0.036 (572 to 752°F)	0.0674 (200°F)

Stainless-steel-UO_2 dispersions are normally fabricated by powder-metallurgy techniques. Uranium dioxide and stainless-steel powder are blended, pressed, and sintered. Final dimensions and further densification are achieved by cold or hot working. The physical and mechanical properties of the dispersions are influenced by many variables including UO_2 particle size, type and amount of UO_2, type of stainless steel, geometry of the element with respect to the property being measured, and fabrication techniques. Reference 18 provides a compilation of mechanical properties.

Zircaloy-UO_2 dispersions have been used as a replacement for the U-Zr alloy used in the first seed of the Shippingport reactor. Such dispersions have greater resistance to radiation damage than the alloy, since damage would be confined to the area around the UO_2 particles rather than being dispersed throughout the alloy.

Normally a dispersion fuel element offers good corrosion resistance regardless of the resistance of fuel particles. Exposure of small amounts of fuel to the coolant due to cladding failure means only a few particles are exposed. If there is stringering of the fuel particles, corrosion can proceed along the interconnected particles. Since UO_2 has good corrosion resistance in high-temperature water, the aqueous corrosion resistance of these dispersions is good even when some stringering occurs; however, the radiation damage resistance of such a fuel would be poor.

Resistance to radiation damage of stainless UO_2 dispersions is good, but as with all fuels, radiation damage places some limits on their performance. In a properly fabricated dispersion fuel, the particles are uniformly dispersed throughout the nonfissile matrix that predominates in volume. Fission products released from the fuel are then confined to narrow regions around the fuel particles; a continuous web of undamaged matrix remains. Radiation-induced swelling arises because of growth of the UO_2 particles due to an accumulation of fission products and the partial escape of fission gases from the UO_2 (Ref. 19). The latter effect is the most significant. A gas-filled void is created around the particles; this pressurizes the matrix shell causing it to expand as a thick-walled vessel under pressure. Thus, the matrix swelling that can be allowed will set the maximum burnup. Swelling limits can be determined on the basis of allowable dimensional changes or a maximum strain determined on the basis of the reduced ductility limit of the neutron-embrittled matrix.

The specific heat of a dispersion can be obtained by combining those of its constituents linearly in accordance with dispersion composition. Densities can be obtained in a similar manner providing an allowance is made for the fact that the UO_2 particles are porous and, therefore, densities of the dispersion range from 92 to 98% theoretical. To obtain the thermal conductivity of a dispersion of small particles of conductivity k_p uniformly dispersed in a continuous substance of conductivity k_s, Jakob[20] recommends an equation originally derived by Maxwell and used for the present purposes by Eucken:

$$k_d = k_s \frac{1 - (1 - ak_p/k_s)b}{1 + (a - 1)b} \, , \tag{2.1}$$

where

k_d = thermal conductivity of dispersion

$a = 3k_s/(2k_s + k_p)$

$b = V_p/(V_s + V_p)$

V_s = total volume of continuous substance

V_p = total volume of distributed particles.

Equation (2.1) was derived for small values of b, but it holds approximately for $b \leqslant 0.5$. We suggest that use above $b \leqslant 0.25$ should be with caution. Stora[21] suggests that for a mixture in which neither phase completely surrounds the other, Bruggeman's equation[22] should be used:

$$\frac{k_s - k_d}{k_s - 2k_d} = \left(\frac{b}{1-b}\right)\left(\frac{k_d - k_p}{2k_d + k_p}\right) . \qquad (2.2)$$

If there is appreciable stringering of the UO_2, thermal conductivity is anisotropic and Eq. (2.1) should not be applied. Under these conditions, Stora[21] suggests that the Fricke equation[23] for dispersion of ellipsoidal particles can be used. He considers that all particles can be reduced to ellipsoids by appropriately varying the relative size of the A, B, and C axes. The anisotropic effect is simulated by changing the size and orientation of the axes of the ellipsoids with respect to the direction of heat flow. Fricke's equation for a dispersion is

$$k_d = k_s\left[\frac{Xk_s - k_p + Xb(k_s - k_p)}{Xk_s + k_p + b(k_s - k_p)}\right] , \qquad (2.3)$$

where

$$X = \frac{(k_p/k_s)(\beta - 1) + 1}{(k_p/k_s) - (\beta + 1)} .$$

When axis A is parallel to the heat flow,

$$\beta = \frac{(k_p/k_s) - 1}{1 + [(k_p/k_s) - 1](1 - M)} , \qquad (2.4)$$

and when axis A is perpendicular to the heat flow,

$$\beta = \frac{(k_p/k_s) - 1}{1 + (k_p/k_s) - (M/2)} . \qquad (2.5)$$

Here, M is a trigonometric function of the A/B length ratio. By using the X parameter, Stora[21] extended Bruggeman's equation for mixtures in which neither phase completely surrounds the other to ellipsoidal particles.

$$\frac{k_s - k_d}{k_s + Xk_d} = \left(\frac{b}{1-b}\right)\left(\frac{k_d - k_p}{Xk_d + k_p}\right) . \tag{2.6}$$

However, his application of this equation to UO_2-metal mixtures, in which the metal fraction $\leqslant 0.3$, yielded k values that differed from measured values by as much as 40%.

2-1.1.3 Uranium Dioxide

This is by far the most popular fuel material for light water reactors (LWRs); uranium dioxide, mixed uranium and plutonium dioxides, and mixed thorium and uranium dioxides are the only ceramic fuels that have received serious attention. Although the carbides can be satisfactory from a radiation damage viewpoint, their reaction with water eliminates them from consideration for LWR service. Uranium dioxide has been found to exhibit excellent dimensional stability to high burnup. It shows appreciable radial cracking,[24] but when properly restrained by fuel element cladding, cracking does not lead to dimensional instability. The generation of fission products leads to a slight swelling of UO_2, which is roughly linear with burnup. Above a critical burnup, the swelling rate increases markedly; this critical burnup can be increased by using slightly less dense UO_2, or by providing more cladding restraint. The critical burnup for unrestrained UO_2 at normal PWR operating conditions is shown in Fig. 2-1 as a function of the void within the fuel. Since the fuel swelling rate also depends on fuel temperature, the critical burnup actually observed depends on the operating power level.

An illustration of the effect of cladding restraint is provided by Dayton,[5] who noted that for platelet-type fuel elements, the increased cladding restraint of $\frac{1}{4}$-in. platelets postponed critical swelling to a burnup over 130% that observed for $\frac{1}{2}$-in.-wide platelets. Sophisticated design calculations now recognize that UO_2 at high temperatures has some plasticity; deformation of the fuel is affected by elastic, plastic, and creep properties of fuel and cladding materials.[25] Thus, Fig. 2-1 can only be regarded as a crude guide to the behavior of an actual rod.

At high flux irradiation, appreciable UO_2 sublimation occurs; this leads to the formation of a central void surrounded by UO_2 of essentially theoretical density. Void migration has been theoretically investigated by De Hales and Horn[26] and Nichols,[27] who derived equations describing the rate of void migration for central temperatures above 2750°F.

At very high power, a molten core is obtained. Although some experimental data indicate such a core is not deleterious,[28] the avoidance of melting was generally accepted as a design criterion. The volumetric expansion of UO_2 on melting can lead to severe swelling of the cladding and, subsequently, to its destruction over a portion of the fuel rod. Failures of this nature were observed by Eichenberg et al.[29] Successful operation with gross central melting was satisfactorily accomplished by Lyons et al.[30] with hollow pellets that provided the free volume necessary to accommodate UO_2 expansion on melting. More recent in-reactor

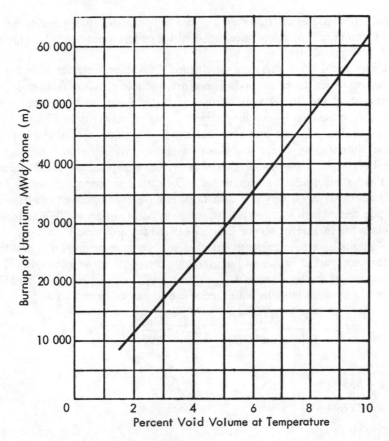

Fig. 2-1.   Critical burnup for rapid swelling of unrestrained UO_2. [From *Nucl. Eng. Des.*, **6**, 301 (1967)].

experiments by Lepescky et al.[31] indicate satisfactory fuel performance with central melting can be obtained with solid pellets.

It has been postulated that under particular conditions of fuel relocation and cracking, molten fuel could come in contact with the cladding, which could be brought to its melting point and burned through. This mechanism may be one of the factors contributing to the 1966 failure in the Plutonium Recycle Test Reactor (PRTR) rupture test loop, where fuel element failure was followed by burn-through of the surrounding pressure tube. The question of whether central melting is a real limit is probably moot since accident analyses now generally lead to the establishment of linear heat ratings considerably below those at which melting occurs.

Current technology provides UO_2 compacts of close to theoretical density by pressed and sintered pellets. As the maximum burnups increased, lower density fuels have been used to reduce swelling. This has led to renewed interest in vibratory compaction, swaging, or combined compaction and swaging of UO_2 powder as production processes. However, the use of pelletized fuel remains the current (1977) commercial practice, since high-density pellets mean less risk of fuel densification during irradiation.

Under usual PWR operating conditions, UO_2 does not react with the coolant and corrosion resistance is good. Therefore, pinhole cladding failures do not lead to washout of the fuel element. However, waterlogging of failed elements at low power and subsequently expelling steam at higher power does lead to the escape of some fission products into the coolant. Under some conditions, it is possible for the waterlogging effect to be severe enough to cause fuel element failure. Eichenberg et al.[32] postulated that this waterlogging failure could occur if the defect were blocked by a particle of UO_2, thus restricting the escape of steam during startup. Only a very few such failures have been observed in a large number of in-pile tests[29] with sintered pellets; no such failure of actual PWR sintered-pellet fuel elements in reactor service has yet (1978) been observed, although defective fuel elements have been present. Waterlogging seems to be more of a problem with vibratory compacted or swaged fuel elements than with those using sintered pellets.

Because of the wide use of UO_2 as a reactor fuel, its properties have been the subject of numerous investigations. Specific heat can be computed as[33]

$$C_p = \frac{K_1 \theta^2 \exp(\theta/T)}{T^2 [\exp(\theta/T) - 1]^2} + 2K_2 T + \frac{K_3 E}{RT^2} \exp(-E/RT) \quad , \tag{2.7}$$

where

$\theta = 535.285$

$E = 37\ 694.6$

$K_1 = 19.1450$

$K_2 = 7.84733 \times 10^{-4}$

$K_3 = 5.64373 \times 10^6$

C_p = specific heat at constant pressure, cal/mol K

T = temperature, K

$R = 1.987$, cal/mol K.

The theoretical density of UO_2 at room temperature is 10.96 g/cm³; its thermal expansion is indicated in Fig. 2-2. The density just below and just above the melting point is reported as 9.65 and 8.80 g/cm³, respectively.[34] Christensen[35] gives the thermal expansion coefficient of molten UO_2 as 3.5×10^{-5} in./in. °C.

Despite the many investigations of UO_2 thermal conductivity, considerable scatter in the data remains. Data of a number of investigators for unirradiated UO_2 are shown in Fig. 2-3, where all have been corrected to 95% theoretical density. The simplest way to correct for density variation is to use the relationship

$$k_{95\%} = k_{measured} \frac{0.95}{1 - \alpha} \quad , \tag{2.8}$$

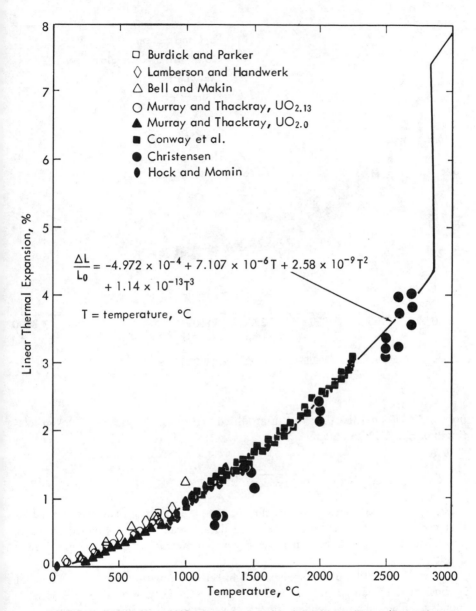

Fig. 2-2.   Comparison of UO_2 thermal expansion data with prediction (Ref. 50).

where α is the void fraction of the sample. Godfrey et al.[36] and May et al.[37] noted that the conductivity of UO_2 increased as the oxygen-uranium ratio is reduced. This stoichiometric effect can account for some of the data scatter. Belle et al.[38] show that data scatter is reduced when a revised method is used for correcting the data to a standard porosity. By assuming that a pellet is a dispersion of nonconducting

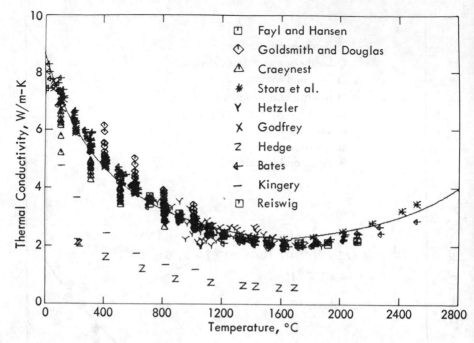

Fig. 2-3.   Thermal conductivity of unirradiated UO_2 (Ref. 50).

pores in a UO_2 matrix, they used a modified version of the Maxwell-Eucken relationship, Eq. (2.1), to obtain

$$k_p = \frac{1 - \alpha}{1 + \beta\alpha} (k_{100\%}) \quad , \qquad (2.9)$$

where k_p is the conductivity of the porous sample in question and β is a constant that depends on the material. A reduced data spread can be obtained by using $\beta = 0.5$ for fuels of 90% theoretical density and above and $\beta = 0.7$ for fuels of lower density.

Stora[21] notes that β corresponds to X in Fricke's equation, Eq. (2.3), if a pellet is considered to consist of a dispersion of pores in UO_2. A β value of 0.5 corresponds to spherical pores, while a larger value corresponds to ellipsoidal pores.

Thermal conductivity data below 2500°F generally follow the expected behavior for lattice conduction. In 1961, on the basis of postirradiation examinations, Bates[39] postulated an increase in conductivity at higher temperatures due to internal thermal radiation. This seems to be borne out by the more recent data of Godfrey et al.[36] and Nishijima[40] for unirradiated UO_2. However, there is disagreement as to whether this is due to thermal irradiation or electron effects. At present, the best estimate of the conductivity of unirradiated UO_2 seems to be a curve drawn through the Godfrey et al. and Nishijima data over the range in which they are available (the solid line in Fig. 2-3). Extrapolation of the curve is in fair agreement with Feith's very high-temperature data.[41]

Considerably more uncertainty is attached to the thermal conductivity of UO_2 after irradiation. In 1959, Runnals[42] noted a decrease in conductivity at low temperatures after an irradiation of 9×10^{17} n/cm^2 and no significant change afterward. Other investigations generally noted qualitatively similar behavior; Fig. 2-4 compares the suggested curve for unirradiated conductivity with the available data for irradiated UO_2. It seems probable that there is some decrease in conductivity below 1500°F and little change between 2000 and 3000°F. The issue is not clear above 3000°F since Bates' estimates[39] do not agree with Lyons' results.[43]

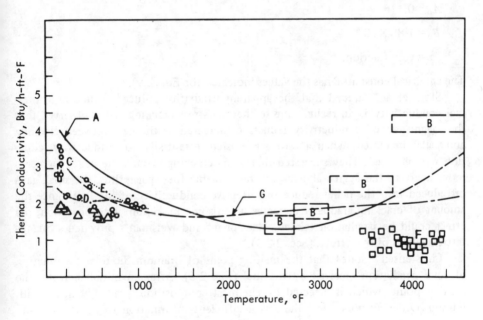

Fig. 2-4.   Irradiated UO_2 thermal conductivity and out-of-pile data. [From *Nucl. Eng. Des.,* **6**, 301 (1967)]. Notation:

A = curve based on unirradiated data

B = from Ref. 39, postirradiation measurements

C = from Bain and Robertson, *J. Nucl. Mater.,* **1**, 109 (1959), postirradiation data

D = Dayton and Tipton, BMI-1448 (Rev.), Battelle Memorial Institute (1960), post-irradiation data (————heating, · · · · · cooling)

E = from Dayton and Tipton, BMI-1448 (Rev.), Battelle Memorial Institute (1960), postirradiation data (Δ first-cycle heating, ○ second-cycle heating, □ third-cycle cooling)

F = from Ref. 43, postirradiation data

G = from Robertson et al., *J. Nucl. Mater.,* **7**, 225 (1962), $\int_{0}^{2800°C} K dt = 97$ W/cm.

In 1967, Belle et al.[38] surveyed all the available data and concluded that conductivity was a function of both burnup and temperature. They proposed the following correlation:

$$k = \left(\frac{1-\alpha}{1+\beta\alpha}\right)[A_0 + B_0 T + (6.23/T) + (49.82/T^2) + 0.0139F(1386/T)]^{-1} , \quad (2.10)$$

where

k = thermal conductivity in Btu/h ft °F

T = °R

F = burnup, fissions/cm^3 X 10^{-20}

A_0 = 0.116

B_0 = 1.88 X 10^{-4}

α = void fraction.

The empirical constant β has the values indicated for Eq. (2.9).

Stora et al.[44] argued that the apparent irradiation-induced reduction in thermal conductivity is, in reality, due to thermal stress cracking. They observed that the magnitude of conductivity reduction increased as the gap between cladding and pellet increased. When a zero gap existed, essentially no conductivity reduction was observed. They suggested that when cracking occurs, the pellet segments move apart to the degree allowed by the available fuel space; thus, the small gaps introduced into the pellet reduce the effective conductivity of the fuel. Since the amount of cracking gradually increases up to some maximum with operation, the effect could be irradiation related. MacDonald and Weisman[45] provided a quantitative analysis of this effect. (Sec. 2-2.5).

Christensen[46] states that the melting point of uranium dioxide is a function of total irradiation received by the fuel (Fig. 2-5). After an initial increase, the melting point, which is affected by chemical composition, gradually drops with burnup. Other authors[47] feel there is insufficient information to come to a conclusion on the change in melting point with irradiation. When less than the stoichiometric amount of oxygen is present, the melting point is reduced. From the data of Guinet et al.,[48] who investigated the entire U-UO$_2$ system, we observe that UO$_{1.8}$ begins to melt at 2690°C in contrast to 2750°C for pure UO$_2$. A brief summary of UO$_2$ properties is provided in Table 2-II and extensive information is provided by Belle[49] and MacDonald and Thompson.[50]

2-1.1.4 Thorium Oxide Fuels

Mixed thorium and uranium dioxides were produced for use as a reactor fuel with the spectral shift control concept. Current interest centers around the use of ThO$_2$ as a blanket in a seed and blanket thermal reactor. The behavior of ThO$_2$ under irradiation is very similar to that of UO$_2$; thermal conductivity is affected by thermal stress cracking of the fuel pellets in the same manner as uranium

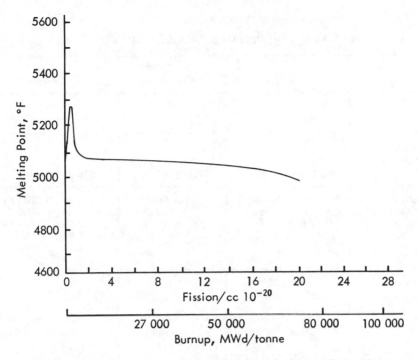

Fig. 2-5.   Melting point of UO_2 as a function of burnup (Ref. 46).

fuel. Therefore, Belle et al.[38] were able to correlate the conductivity of ThO_2 and mixed $ThO_2 + UO_2$ by Eq. (2.10) using the constants of Table 2-III.

In-pile thermal conductivities of ThO_2-PuO_2 mixtures containing 1.1 to 2.7 wt% PuO_2 were measured by Jeffs,[51] who observed conductivities ~25% below those obtained by laboratory measurements of pure ThO_2. Within the scatter of the data, he saw no difference between the compositions tested.

Godfrey et al.[52] measured the specific heat of ThO_2 over a wide range. These and other thermal properties are summarized in Table 2-II.

2-1.1.5 Mixed Plutonium and Uranium Oxides

Primary interest in mixed UO_2-PuO_2 fuels stems from their application to fast breeder technology. However, before development of a fast breeder market, there is interest in recycling plutonium back into the water-cooled reactors from which it is generated. Available information indicates that mixtures of PuO_2 and UO_2 provide a satisfactory water reactor fuel. As expected on the basis of the similarity of the two compounds, there is little difference in the behavior of mixed oxides and UO_2. In view of the small amounts of PuO_2 that would be present in PWR fuel (<6%), it is generally satisfactory to use UO_2 properties as an approximation of the mixed oxides.

TABLE 2-II

THERMAL PROPERTIES OF CERAMIC FUELS

Fuel	UO_2	UO_2–80 at.% PuO_2–20 at.%	ThO_2
Fuel density (g/cm^3)	10.97	11.08	10.01
Melting temperature $(°F)$	4980 (see Fig. 2-5)	5036	5970
Thermal conductivity $[Btu/(h\ ft\ °F)]$	2.54 (930°F) 1.04 (3630°F) (see Fig. 2-3)	2.02 (930°F) 1.04 (3630°F)	7.29 (200°F) 5.34 (400°F) 3.59 (700°F) 2.68 (1000°F) 2.07 (1400°F) 1.68 (2400°F)
Thermal expansion $(in./in.\ °F)$	6.12×10^{-6} (75 to 5070°F) (see Fig. 2-2)	6.12×10^{-6} (75 to 5035°F)	
Heat capacity $(Btu/lb\ °F)$	0.0567 (90°F) 0.0755 (135°F) 0.090 (3150°F) 0.118 (4050°F)	approximately the same as UO_2	0.0547 (90°F) 0.0696 (1350°F) 0.0775 (3150°F) 0.082 (4050°F)

TABLE 2-III

CONSTANTS (Ref. 38) FOR THERMAL CONDUCTIVITY CORRELATION

Material	A_0	B_0
ThO_2	0	2.21×10^{-4}
ThO_2 + 10 wt% UO_2	0.0225	1.78×10^{-4}

MacDonald and Thompson[50] correlated the linear thermal expansion of PuO_2 by

$$\frac{\Delta L}{L} = -3.9735 \times 10^{-4} + 8.4955 \times 10^{-6}T + 2.1513 \times 10^{-9}T^2 + 3.7143 \times 10^{-16}T^3 \ ,$$

(2.11)

where T is in °C. They suggest the expansion of mixtures of UO_2 and PuO_2 can be calculated from the weighted average of the expansion of UO_2 and PuO_2.

Epstein[53] measured the melting points of UO_2-PuO_2 mixtures and observed melting points of 2840 ± 20°C for UO_2 and 2390 ± 20°C for PuO_2. His data show a typical binary pair without a maximum or minimum in the liquidus or solidus curves and, for small additions of PuO_2, his results can be closely approximated by assuming a linear decrease in melting point per atom percent of PuO_2 addition. Krankota and Craig[54] observed that the melting point of mixed oxides decreases slightly with burnup. After 85 000 MWd/tonne burnup, the melting point of $(Pu_{0.25}U_{0.8})O_2$ decreased 50°C.

The thermal conductivity of PuO_2 was measured by Lagedrost et al.[55] These measurements corrected to 100% density show a decrease in conductivity from 0.06 W/cm °C at 300°C to 0.025 W/cm °C at 1200°C. Due to interest in the mixture as a fast reactor fuel, the thermal conductivity of $(U_{0.8}Pu_{0.2})O_2$ was measured by a number of laboratories.[56,57] Biancheria et al.[58] give the thermal conductivity of 95% dense $(U_{0.8}Pu_{0.2})O_2$ as

$$k = [3.11 + 0.272T(K)]^{-1} + 5.39 \times 10^{-13} T^3(K) \quad ,$$

where k is in W/cm °C. MacDonald and Thompson[50] state that for plutonium contents between 5 and 20%, the plutonium content has little influence on k for temperatures up to 1600°C. Between 1000 and 1900°C, the $(U_{0.8}Pu_{0.2})O_2$ of 95% density exhibits thermal conductivities very similar to UO_2 of the same density. Below 1000°C, the thermal conductivity of UO_2 is somewhat higher. As expected on the basis of UO_2 behavior, a stoichiometry effect on conductivity is shown by $(U_{0.8}Pu_{0.2})$ mixed oxides. Hetzler et al.[57] found that hypostoichiometric oxides $(Pu - U)O_{1.98}$ had somewhat lower conductivities than the stoichiometric material. Significant thermal design properties for $(U_{0.8}Pu_{0.2})O_2$ are summarized in Table 2-IV. Properties of other mixtures can be estimated by interpolation.

2-1.2 Fuel Element Cladding and Assembly Designs

2-1.2.1 Cladding Properties

Although the most common PWR fuels have good corrosion resistance, continued exposure of any of these to the coolant would result in coolant activity levels far beyond those consistent with direct maintenance. Therefore, it is the universal practice to clad various fuels by a thin layer of corrosion-resistant metal. Ideally, the cladding should be inert to the coolant, have a high strength and ductility that are unaffected by radiation, have a low-neutron-absorption cross section, and be fabricated easily and economically. Since no such material is known, one must sacrifice some attributes to gain those believed most important.

Modern PWRs use stainless-steel-clad or Zircaloy-alloy-clad fuel exclusively. Low neutron cross section, excellent corrosion resistance to low-temperature water, and low cost made aluminum an early candidate for this use. Unfortunately, common aluminum alloys do not have the necessary corrosion resistance in high-temperature water to allow them to be considered. Argonne National Laboratory (ANL) conducted extensive investigations on various aluminum alloys containing small amounts of nickel (i.e., 1%) and lesser amounts of iron and other constituents.

TABLE 2-IV

PROPERTIES OF CLADDING MATERIALS

Material	Zircaloy-2	347 Stainless Steel
Composition	1.2 to 1.7% tin 0.07 to 0.2% iron 0.05 to 0.15% chromium 0.03 to 0.06% nickel balance zirconium	17 to 19% chromium 9 to 12% nickel 0.8% columbium 0.2% max manganese 0.08% max carbon 1% max silicon
Density (g/cm^3)	6.57 (room temperature)	8.03 (room temperature)
Melting point	3360°F	2550 to 2600°F
Thermal conductivity [Btu/(h ft °F)]	6.82 (100°F) 6.89 (200°F) 7.11 (400°F) 7.37 (600°F) 7.64 (800°F) 7.77 (900°F)	8.6 (100°F) 9.0 (200°F) 9.8 (400°F) 10.6 (600°F) 11.5 (800°F) 12.4 (1000°F)
Mean coefficient of thermal expansion (in./in. °F)	4.62×10^{-6} (25 to 800°C) (rolling direction) 6.85×10^{-6} (25 to 800°C) (transverse direction)	9.05×10^{-6} (68 to 100°F) 9.25×10^{-6} (68 to 200°F) 9.55×10^{-6} (68 to 400°F) 9.8×10^{-6} (68 to 600°F) 10.0×10^{-6} (68 to 800°F) 10.25×10^{-6} (68 to 1000°F)
Heat capacity [Btu/(lb °F)]	0.0725 (200°F) 0.0764 (400°F) 0.0789 (600°F) 0.081 (800°F) 0.0831 (1000°F) 0.0853 (1200°F)	0.12 (32 to 212°F)

Although these alloys have significantly reduced corrosion rates, ANL found the rates are still excessive at PWR conditions.[59]

Stainless Steel. With excellent corrosion resistance and high strength, the 300-series stainless steels possess many of the properties of an ideal cladding. Although their properties are slightly affected by radiation (yield point increases and ductility decreases appreciably), their radiation resistance is good, and they are readily fabricated without special techniques. Their major drawback is a relatively high neutron cross section, which requires an increase in initial enrichment. These alloys are not suitable for reactors where natural uranium is the fuel.

Stainless steel is known to be subject to stress-assisted corrosion and stress corrosion cracking in high-temperature water when oxygen and halogens are present. Duncan et al.[60] reported stress-assisted corrosion failures of this nature in the cold-worked stainless cladding of fuel in BWR fuel. No such cracking has been reported in PWRs, perhaps because of the lower oxygen content of the coolant. However, it presents a severe problem in superheat systems, where experience has been poor, and would be troublesome in designing a supercritical PWR.

The stainless alloy chosen for a particular case depends on the fabrication techniques used. All of the 300-series alloys have very nearly the same thermal design properties. Typical values are presented in Table 2-IV.

Zirconium Alloys. Hafnium-free zirconium has a very low-absorption cross section (0.18 b compared to 2.43 b for iron and 4.5 for nickel) and good resistance to water corrosion at high temperatures. The addition of small amounts of tin and iron significantly improves the strength of zirconium. As a result of extensive investigations at the Bettis Atomic Power Laboratory, three alloys have been developed that have the required strength and corrosion resistance—Zircaloy-2, -3, and -4.

Zircaloy is now used as fuel element cladding in all light-water-cooled reactors built for central power production and Zircaloy-4 (1.3% tin, 0.22% iron, 0.10% chromium, balance zirconium) is generally used for PWR fuel cladding.

Although the melting point of zirconium is high (1852°C), a phase change at 862°C affects the mechanical properties. At this temperature, pure zirconium goes from a close-packed hexagonal structure, α, to one that is bcc.

Zircaloy tubing or plate is more expensive than stainless steel, and fabrication must be more carefully controlled. Oxygen must be excluded from all welds; welding must be done in an inert atmosphere, in a glove box or similar enclosure. Since zirconium alloys creep at relatively low stresses at PWR temperatures, allowance must be made in the mechanical design of the fuel elements. However, creep can be beneficial since it acts to relieve any high-cladding stress. Ibrahim[61] correlated Zircaloy creep by

$$\epsilon = [2.768 \times 10^{-5} \exp(1.5 \times 10^{-2}T + 9.663 \times 10^{-5}\sigma)] [t^{(1.655 \times 10^{-3}T - 0.297)}] \quad,$$

$$(2.12)$$

where

ϵ = strain, %

T = temperature, °C

t = time, h

σ = effective stress, psi.

The rate of creep is accelerated by irradiation; in-pile creep rates can be several times those obtained in out-of-pile tests.

Zirconium and its alloys react significantly with steam at high temperatures $[Zr + 2H_2O(g) \rightarrow ZrO_2 + 2H_2(g)]$. The energy release due to this reaction must be considered when accidents that can give rise to high-cladding temperatures are analyzed. At temperatures above 500°C, the hydrogen produced and the oxygen of the oxide layer diffuse into the metallic phase, thereby reducing the ductility of the metal. Stainless-steel cladding at high temperatures also reacts chemically with steam. However, there the reaction rate does not become appreciable until temperatures in excess of 1000°C are reached.

If interfacial temperatures rise above 675°C, Zircaloy cladding can react with UO_2. Since such temperatures are <400°C in normal operation, the reaction needs to be considered only for accident situations.

At normal operating temperatures, zirconium reacts slowly with water to form zirconium oxide and hydrogen. At operating conditions, oxygen from the oxide layer does not diffuse into the metal, but hydrogen produced partly diffuses through the oxide layer into the metal. The quantity of hydrogen (given as a fraction of the total hydrogen formed) that diffuses into the metal is called the "pickup fraction." Zircaloy-4 has a pickup fraction between 0.005 and 0.2 (Ref. 62).

In-reactor corrosion of Zircaloy is significantly increased by fast neutron flux under oxygenated coolant conditions. However, under low-oxygen conditions, fast neutron flux increases the corrosion rate only slightly. This latter situation generally holds in a PWR since hydrogen overpressure in the pressurizer inhibits radiolysis. In the Saxton core,[63] weight gains on the order of 200 mg/dm² were observed after 500 full power days at 339°C. The total corrosion can be related to the average hydrogen concentration, C_H, in Zircaloy by using

$$C_H(\text{ppm}) = 1.25 \times 10^{-2} \alpha' \frac{\Delta w}{d\rho_{Zr}} , \qquad (2.13)$$

where

α' = pickup fraction

Δw = total weight gain, mg/dm²

d = cladding wall thickness, cm

ρ_{Zr} = zirconium density, g/cm³.

Hydriding of Zircaloy cladding can also originate from the internal cladding surface. In the presence of water contamination of UO_2 pellets, internal hydriding can lead to bulges or blisters on the cladding surface due to the 13% volume change in the $Zr \rightarrow ZrH_{1.6}$ reaction. Internal hydriding has been one of the most persistent sources of fuel rod failure. The attack is localized and seems to occur at local hot spots or regions where the zirconium oxide film has been damaged. To avoid such attack, concentration of water vapor in the gas inside a fuel rod must be kept below 2 mg/cm³ (Ref. 64). Manufacturers generally specify that the water contents of UO_2 pellets be held at a few ppm so this critical concentration does not arise. Hydrogen trapped in pellet pores is sometimes removed by getters.

Fission products released from UO_2 can react chemically with zirconium. Iodine-induced stress corrosion cracking can occur when locally high stresses arise, such as during power ramps.[65] The possibility of such brittle cracking means that considerable care must be exercised in the rate at which the power of any fuel assembly is increased. High local stresses can be generated in the cladding of CANDU reactor fuel elements when power ramping accompanies axial fuel shuffling. High-burnup fuel rods in vessel-type reactors can also experience high-cladding stresses during rapid load changes or on startup after refueling. A number of fuel failures have been attributed to this.[66]

Irradiation of zirconium and its alloys increases their hardness and decreases the ductility. This decrease in ductility must be considered in designing reactor components such as pressure tubes. In fuel element design, it often leads to a limitation on the maximum allowable cladding strain.

Thermal and mechanical properties of Zircaloy-2 are presented in Table 2-IV.

High Nickel Alloys. Sensitivity of the 300-series stainless steels to stress-corrosion cracking under boiling and superheat conditions, as well as the requirement for strength at high temperatures, has led to the consideration of high-nickel alloys for supercritical reactors. Various Inconel and Incoloy compositions (up to 72% nickel as contrasted to 8% nickel for 300-series stainless) have been proposed. These alloys are resistant to stress-corrosion cracking and have good high-temperature strength. Allio and Thomas[67] reported that Westinghouse 16–20 alloys (16% chromium and 20% nickel with low carbon and nitrogen) behaved satisfactorily at high fluxes in an environment of supercritical water. These alloys have a macroscopic neutron absorption cross section of 0.266/cm as compared to 0.300/cm for Incoloy. Reference 50 presents the design properties of a number of high-nickel alloys.

2-1.2.2 Fuel Assembly Designs

Rods. Metallic, low-enrichment uranium fuel has almost always been used in the form of clad, solid or annular, cylindrical rods. Tubing for the cladding is commercially available, fabrication is relatively simple, and the necessary surface area is obtained by proper selection of rod size. When aluminum is used as the cladding, a diffusion bond between the cladding and fuel is usually formed by a thin layer of aluminum-silicon alloy. Zircaloy-clad elements were formed for the Hanford New Production Reactor (NPR) by coextrusion of a copper-zirconium-uranium billet, with the copper stripped away after extrusion.

Rod elements are also most commonly used for uranium dioxide fuel; they are generally formed by loading pressed and sintered UO_2 pellets into Zircaloy or stainless-steel tubes to which end plugs are welded. Individual rods are assembled into bundles and end plates, and fittings are affixed. Figure 2-6 shows one of the earlier assemblies for a pressurized water design where brazed ferrules hold the rods together. In more recent designs (Fig. 1-4), the rods have been positioned by several grids containing spring fingers. This design allows higher strength cold-worked cladding to be used, which reduces the amount of absorber in the core and allows the fuel rods to expand, which reduces thermal stresses.

Fuel assemblies designed for the CANDU pressure tube reactor are considerably different from those used in vessel-type PWRs. Since such an assembly must

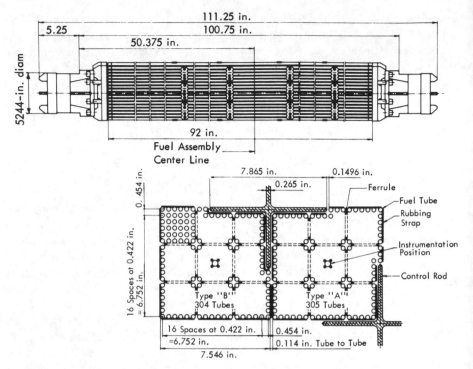

Fig. 2-6.   Original Yankee fuel assembly. [From *Directory of Nuclear Reactors,* Vol. IV, p. 34, International Atomic Energy Agency, Vienna (1962).]

fit within a pressure tube, it contains only a small number—19 or 28—of fuel rods. These assemblies have all been short with bundle lengths no more than six times the diameter of the rod cluster. In one design, a completely integral structure with end plates welded to the ends of all the pins is provided.[65] Spacing between the fuel rods is maintained by metal warts brazed obliquely to the fuel pins. In other designs for pressure tube assemblies, wire wraps, ferrules, and spacer pads have been used to maintain rod spacing. A drawing of a typical CANDU fuel assembly is shown in Fig. 2-7.

Plate-Type Elements. Designs for highly enriched uranium-Zircaloy elements and UO_2-stainless-steel dispersion elements are similar. In both cases, the so-called "picture frame" method is used for fabrication, which means that a sheet of the fuel alloy, or dispersion, is rimmed by a frame of cladding. Upper and lower sheets of cladding are then welded to the frame and the pack is rolled to final dimensions. A group of fuel plates are then placed in a box-like enclosure made of cladding material. This maintains plate spacing and provides an assembly that can be handled. Plate-type elements containing uranium-Zircaloy fuel used in Shippingport Core I are illustrated in Fig. 2-8. Plate-type elements using a UO_2-stainless-steel dispersion as fuel have been used in the Army Package Power Reactor (APPR).

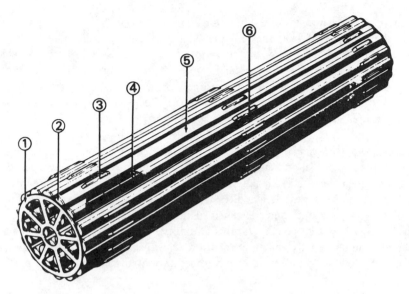

Fig. 2-7.   CANDU fuel element rotation: ① Zircaloy structural end plate, ② Zircaloy end cap, ③ Zircaloy bearing pads, ④ uranium dioxide pellets, ⑤ Zircaloy fuel sheath, ⑥ Zircaloy spacers. [From J. A. L. Robertson, AECL-4520, Atomic Energy of Canada Ltd. (1973).]

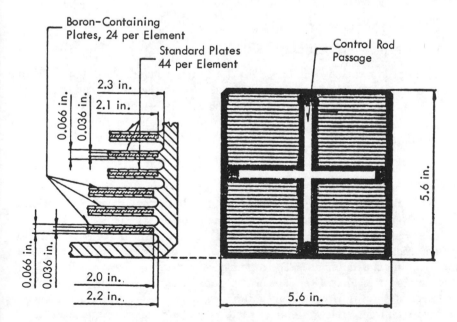

Fig. 2-8.   Shippingport Core 1 seed assembly. [From *Directory of Nuclear Reactors,* Vol. IV, p. 21, International Atomic Energy Agency, Vienna (1962).]

Plate-type elements for UO_2 fuel have been used in Shippingport Core II. Rectangular platelets of UO_2 fuel are placed in a slotted Zircaloy plate, which is sandwiched between two Zircaloy cover plates. The fairly substantial Zircaloy plates serve to restrain the UO_2 platelets under high burnup. Such plates can be fabricated by the gas-pressure bonding process[5] in which the fuel element can be assembled from plates and strips that are subsequently bonded together at high pressures and temperatures.

Tubular Fuel Elements. Tubular fuel elements can be fabricated from sandwiches of dispersion fuel and cladding. Standard stainless-steel forming operations can be used to make the sandwich sheets.[68] Elements of this type were used in portable reactors (PM series) designed for the U.S. Army nuclear power program.

Annular fuel elements, such as those of the first USSR power station, can be formed from hollow UO_2 pellets. Various vibratory compaction techniques can also be used for annular fuel element fabrication. Multipass, concentric ring, annular elements in individual pressure tubes have been proposed for supercritical pressure reactors.[69]

2-2 BEHAVIOR OF UO_2 FUEL ELEMENTS

Since fuel elements consisting of UO_2 pellets in cylindrical Zircaloy tubes are used in nearly all power reactors, the special design properties of these elements are examined.

2-2.1 Mechanical Properties of UO_2

At temperatures below $\sim 1000°C$, UO_2 is a brittle material. Below the brittle to ductile transition temperature, T_c, fracture stress is smaller than yield stress and brittle fracture occurs. Hence, at relatively modest thermal stresses, cracking of a UO_2 pellet occurs. Below T_c, the stress-strain curve is linear and Young's modulus, E, can be obtained from[70]

$$E = 22.9 \times 10^4 - 20.1T - 58.7 \times 10^4 \alpha \ , \qquad (2.14)$$

where E is in N/mm^2, T is in K, and porosity α is given as a fraction. The value of E depends slightly on the oxygen-uranium ratio. If the oxygen-uranium ratio is increased from 2.0 to 2.16, an $\sim 22\%$ decrease in E can be expected.[70]

The typical value of Poisson's ratio, γ, is reported to be 0.316 (Ref. 70). No data are available on temperature dependence, but other ceramic materials show little effect of temperature on γ. The dependence on porosity is very small.

Fracture stress σ_f of UO_2 depends on porosity, grain size, and temperature. The effect of porosity is small[70] and can be neglected. Both temperature and grain size have significant effects, although there seems to be some disagreement as to the magnitude of these effects. Stehle et al.[70] suggest that fracture stress can be expressed by

$$\sigma_f = \sigma_{f_0} + AT \quad , \tag{2.15}$$

where σ_f is in N/mm^2 and T is in K. Evans and Davidge[71] and Canon et al.[72] indicate that σ_{f_0} varies from ~98.5 to 151 and A from 0.025 to 0.028.

The brittle to ductile transition temperature, T_c, is approximately equal to about half the absolute melting temperature. The value of T_c also varies with porosity, grain size, and strain rate. At zero or low strain rates, T_c varies from ~1000 to 1100°C. Increasing the porosity of the pellet slightly decreases T_c.

A second transition temperature, T_t, is defined as the temperature that shows a marked decrease in the ultimate tensile stress measured in bending tests. In the intermediate range, $T_c < T < T_t$, some deformation occurs before fracture (semi-brittle fracture). For $T > T_t$, the fracture mode is completely ductile. The value of T_t depends on strain rate and porosity. At low strain rates, T_t is on the order of 1400°C, while at very high strain rates it can increase to 1800°C (Ref. 72).

At high temperatures, UO_2 exhibits significant thermal creep. However, fissioning oxide fuels exhibit significantly enhanced plasticity and show fission-induced creep at temperatures where thermal creep does not occur. Olsen[73] presented an empirical correlation of the data on thermal creep and combined thermal and fission-induced or enhanced creep. He proposes

$$\dot{\epsilon} = \frac{(A_1 - A_2 \dot{F})\exp(-Q_1/RT)}{(A_3 + d)G^2} + \frac{A_4 \sigma^{4.5} \exp(-Q_2/RT)}{(A_5 + d)} + A_6 \sigma \dot{F} \exp(-Q_3/RT) \quad ,$$

$$\tag{2.16}$$

where

$\dot{\epsilon}$ = creep rate, in./in. h

σ = compressive stress, psi

$\dot{F}$ = fission rate, fissions/m^3 s

G = grain size, μm

R = gas constant, 1.987 cal/g mole K

T = temperature, K

$A_1 = 9.728 \times 10^6$

$A_2 = 3.24 \times 10^{-12}$

$A_3 = -87.7$

$A_4 = 1.376 \times 10^{-4}$

$A_5 = -90.5$

$A_6 = 9.24 \times 10^{-28}$

$Q_1 = 90\,000$ cal/mole

$Q_2 = 132\,000$ cal/mole

$Q_3 = 5200$ cal/mole

d = theoretical density, %.

When stressed, plastic flow of UO_2 can lead to densification of fuel pellets through "hot hydrostatic pressing." Stresses on the UO_2 can arise when fuel swelling brings the outer diameter of the fuel pellet in contact with the cladding. Such stresses can also arise if pellet jamming at some elevation prevents free axial expansion of the stack. In contrast to creep, which is governed by deviatoric stress, hot pressing is governed by hydrostatic pressure (Sec. 2-4.1). Sun and Okrent[74] indicate that the volumetric strain, ϵ_{hp}, under hydrostatic compression can be expressed by

$$\epsilon_{hp} = \frac{C}{T}\exp(-Q/T)\sigma\left(1 - \frac{\rho}{\rho_{th}} - \epsilon_{sw}\right)\Delta t \quad , \tag{2.17}$$

where

σ = hydrostatic stress $= (\frac{1}{3})(\sigma_r + \sigma_\theta + \sigma_z)$, dyn/cm^2

$C = 4.7 \times 10^6$ K cm^2/s dyn

$Q = 4.43 \times 10^4$ K

T = absolute temperature, K

ρ_{th} = theoretical density of fuel

ρ = fuel density

ϵ_{sw} = fission gas fuel swelling strain.

It is assumed that hot pressing strain components are all equal

$$[(\epsilon_{hp})_\theta = (\epsilon_{hp})_r = (\epsilon_{hp})_z] = (\tfrac{1}{3})(\epsilon_{hp}) \quad .$$

2-2.2 Fuel Densification and Restructuring

Hot pressing is not the only mechanism that can lead to in-pile densification of UO_2. Pellet densification and shrinkage can be caused by irradiation-induced dissolution of pellet porosity. Many investigators believe densification results from dispersing fine pores as vacancies by fission fragments. These vacancies subsequently diffuse to a grain boundary leaving a densified fuel grain.[75,76] The process is illustrated in Fig. 2-9.

Irradiation-induced fuel densification is of greatest importance in low-density (e.g., ≤92% theoretical) fuels. When low-density fuels were first used, irradiation-induced densification caused pellet shrinkage that led to gaps in the pellet stack and cladding collapse in the space between pellets. The problem is now understood; reactor vendors control densification by limiting the initial volume of fine pores in the fuel.[76] Increased sintering temperature and sintering time lead to a significant decrease in densification.

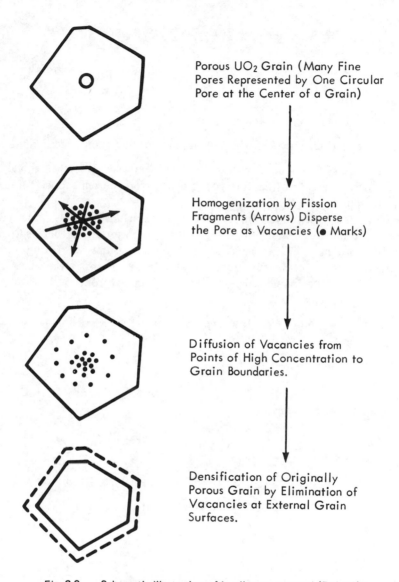

Porous UO₂ Grain (Many Fine
Pores Represented by One Circular
Pore at the Center of a Grain)

Homogenization by Fission
Fragments (Arrows) Disperse
the Pore as Vacancies (● Marks)

Diffusion of Vacancies from
Points of High Concentration to
Grain Boundaries.

Densification of Originally
Porous Grain by Elimination of
Vacancies at External Grain
Surfaces.

Fig. 2-9.   Schematic illustration of in-pile pore removal (Ref. 76).

Several approaches have been suggested for estimating the rate of densification. Rolstad et al.[77] suggested a relationship between percent densification, sintering temperature (a measure of initial pore and grain size), initial density, and burnup. The graphs required for the use of this model are reproduced in Ref. 78.

Stehle et al.[70] point out that the net densification observed is the algebraic sum of fuel swelling and densification. On the basis of kinetics of vacancy generation, they suggest that pellet densification can be represented by

$$V/V_0 = K_1(BU) - K_2\{1 - \exp[-K_3(BU)]\} - \sum_i K_i\{1 - [1 - K_i(BU)]^3\} \ , \quad (2.18)$$

where

BU = burnup

V/V_0 = fraction change in volume

K = empirical constraints that depend on pore and grain size distribution.

The first term of the preceding equation represents swelling; the second term represents disappearance of the fine pore fraction, and the remaining terms represent the disappearance of pores in the micron size range.

Marlowe[79] suggested an alternative theory to explain irradiation-induced densification. He proposes an irradiation-enhanced diffusion model. It follows Coble's[80] theory for sintering, which is based on the assumption that the driving force for pore shrinkage is the effort to reduce free energy by lowering pore surface area. According to Marlowe, in-pile pellet density, ρ, is obtained from

$$\rho = \rho_0 \exp(-S\dot{F}t) + \frac{M}{A} \exp\left[-S\left(\frac{G_0^3}{AD_{irr}} + \dot{F}t\right)\right]\ln\left(1 + \frac{A'D_{irr}}{G_0^3}\dot{F}t\right) \ , \quad (2.19)$$

where

ρ_0 = initial density

A = grain growth constant

$\dot{F}$ = fission rate

M = densification rate constant

D_{irr} = constant relating irradiation-induced diffusivity to fission rate

S = volumetric swelling rate for 100% dense fuel

t = irradiation time

G_0 = initial grain size.

Quantities M and A can be determined from experiments on the thermal sintering of the pellets under study. Buescher and Horn[81] suggest that D_{irr} can also be calculated from thermal sintering data by using $D_{irr}/D' = 1.15 \times 10^{-15}$ cm^3 s/fission, where D' is the thermally activated bulk diffusion rate.

Thermal effects also lead to in-pile restructuring of UO_2. In Sec. 2-1.1, it was noted that porous oxide fuel, irradiated at high heat fluxes, shows large columnar grains distributed around a cylindrical central void. In out-of-pile experiments, MacEwan and Lawson[82] showed that columnar grains would be produced at temperatures above 1700°C if a sufficiently high-temperature gradient were maintained. Columnar grains resulted from the migration of voids, transversely oriented with respect to newly formed grains, toward high-temperature regions.

The observed grain growth is attributed to a vaporization-condensation process that results in a net movement of the pore at high temperatures. We assume that fuel vaporization occurs at the high-temperature end of the pore and condensation occurs at the low-temperature end. Nichols[27] developed a theoretical model for this process. The velocity of a spherical pore, which Nichols shows to be independent of its size, is given by

$$v = \frac{qP_0 \Omega \Delta H_v [N_a(M_1 + M_2)]^{1/2} \exp[-\Delta H_v/(RT)]}{16 P l^2 T^{3/2} (2\pi K M_1 M_2)^{1/2}} \frac{dT}{dr} \quad , \tag{2.20}$$

where

P_0 = constant in vapor pressure equation

$$P = P_0 \exp[-\Delta H_v/(RT)] \quad ,$$

where P_0 is 1.64×10^{14} dyn/cm^2 for UO_2

Ω = atomic volume, 3×10^{-23} cm^3 for UO_2

ΔH_v = heat of vaporization, $\approx 142\,600$ cal/mole for UO_2

M_1 = gram molecular weight of matrix, g

M_2 = gram molecular weight of vapor, g

P = pressure of gas vapor in pore, dyn/cm^2

N_a = Avogadro's number, 6.023×10^{23}/mole

l = cross-sectional radius for collisions between matrix and vapor molecules, 3×10^{-8} cm for UO_2

K = Boltzman's constant, 1.38×10^{-16} erg/°C

r = pellet radius, cm

R = Ideal Gas Law constant, cal/g mole K

T = temperature, K.

Nichols takes the region in which all pores have had sufficient time to migrate to the central void as the columnar grain growth region. Using Eq. (2.20), we can compute the radial boundary of the grain growth region, r_{cg}. Figure 2-10 presents the results of Nichols' calculations for a typical UO_2 pellet. He assumed the surface temperature of the pellet was 800 K and that a parabolic temperature gradient existed across the fuel. Helium was assumed to be the primary vapor constituent. Observe that r_{cg} can become a significant fraction of r_0, the total pellet radius, at high temperatures. The radius of the central void, r_v, is computed by

$$(r_{cg}/r_0)\alpha^{1/2} = r_v/r_0 \quad , \tag{2.21}$$

where α is the initial void fraction of the pellet.

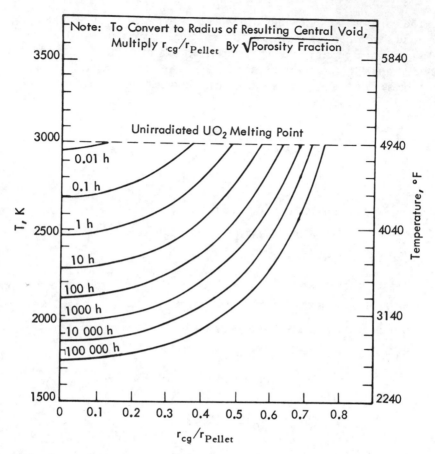

Fig. 2-10.   Columnar grain growth radius in UO₂ rods. [From *J. Nucl. Mater.,* **22**, 214 (1967).]

Nichols discusses the possible instability of the spherical pore shape and the possibility of void transformation into an elongated form in a subsequent paper.[83]

2-2.3 Fuel Swelling

Previously we observed that uranium dioxide swells during irradiation. The data indicate that the rate of fuel swelling is a function of temperature. The swelling initially fills the pores of the fuel. When the portion of the porosity that is effective in accommodating swelling has been consumed, net dimensional changes occur. Table 2-V indicates approximate swelling rates and effective porosities for several temperature ranges.[84] The burnup at which the effective porosity is consumed is similar in concept to the critical burnup for rapid swelling (see Fig. 2-1).

Godesar et al.[85] suggest an analytical expression for total swelling, which was calibrated to a series of in-reactor experiments. They propose

TABLE 2-V

APPROXIMATE SWELLING RATES AND EFFECTIVE POROSITIES

Temperature Range	Fuel Swelling % ($\Delta V/V$) per 10^4 MWd/tonne	Effective Component of Initial Voidage (%)
<1300°C	1.6	30
1300 to 1700°C	1.7	50
>1700°C	0.7	80

$$\Delta V/V_0 = f\alpha\left[\exp\left(-F\frac{\Delta V}{V_0}\right) - 1\right] + \beta(1-\alpha)b \quad , \tag{2.22}$$

where

$\Delta V/V_0$ = relative volume expansion by swelling

b = burnup, MWd/1000 kg U

α = initial porosity

β = swelling rate of 100% dense oxide

  = 1.6×10^{-6} (MWd/1000 kg U)$^{-1}$ for usual PWR fuel temperatures

f = density correction factor = 0.149 $(\alpha)^{-0.746}$

F = geometrical factor = 100.

This model takes into account the influence of porosity on swelling rate. As the volume of pores decreases with time, due to inner swelling, the apparent outer swelling rate increases toward an asymptotic value.

The most recent approaches to fuel swelling have been more mechanistic. Fuel swelling is generally attributed to production of bubbles by gaseous fission products and the accumulation of solid fission products. Solid fission products produce swelling of between 0.5 and 1% per at.% burnup.[86,87] The contribution of solid fission products to swelling is most important at low temperatures.

At high temperatures, gaseous fission products are the major cause of fuel swelling. By calculating the volume of gas bubbles within the fuel grain, the increase in fuel volume and, hence, swelling, can be estimated. Several models have been proposed for this purpose. Nichols and Warner's BUBL-1 code[88,89] calculates the nucleation of gas bubbles, their coalescence, interaction with grain boundaries and dislocations, and migration of bubbles due to the fuel temperature gradient. The model assumes that

1. bubble sizes are determined by gas pressure balancing surface tension and hydrostatic pressure

2. there is a critical force required to pull a bubble away from a dislocation or grain boundary

3. bubble migration along the temperature gradient is due to a surface diffusion mechanism.

Fission gas behavior is followed by a Monte Carlo technique. When two bubbles overlap over a dislocation, they are assumed to coalesce. When a bubble grows sufficiently, the driving force due to the temperature gradient pulls it off the dislocation and the bubble migrates to a grain boundary by surface diffusion. The bubble remains at the grain boundary until further coalescence allows it to grow large enough to be pulled away. The bubble then migrates along the temperature gradient. Release of the gas is assumed to occur after the bubble has gone a distance approximating the distance to a crack or surface.

Yu-Li et al.[90] describe a somewhat similar model. However, their approach, which is embodied in the GRASS code, is somewhat more general. The general approach of GRASS is basically the same as BUBL in that coalescence, release from defects and boundaries, and migration along the temperature gradient are followed. However, the model allows for radiation-induced resolution of the bubbles as a negative term in bubble growth. The GRASS code also allows for random motion of bubbles and coalescence during migration. Surface diffusion is not the sole method for bubble diffusion. Diffusion components are calculated on the basis of surface diffusion, volume diffusion, and evaporation-condensation. In each fuel region and for each bubble size, the mechanism giving the highest bubble velocity is chosen.

Net swelling is seen as the algebraic sum of irradiation-induced densification and fission product swelling

$$(\Delta V/V_0)_{net} = (\Delta V/V_0)_{swell} - (\Delta V/V_0)_{densification} \quad . \tag{2.23}$$

Therefore, a small amount of fuel densification may not be injurious since it may partly counteract the effects of swelling.

2-2.4 Fission Gas Release and Internal Pressure

A significant fraction of the gases produced by fission is released by UO_2. It is generally agreed that below ~600 to 800°C, the rate of fission gas release is controlled by recoil and knockout.[91,92] Release rates are low and not temperature dependent.

In the region between 800 and 1800°C, fission gas release rates are appreciable and markedly temperature dependent. For a number of years, the rate of gas release in this region was believed to be determined by the rate at which gas diffused out through solid UO_2 particles. Available data were correlated using a fission gas diffusivity, D, which followed an Arrhenius rate equation

$$D = D_0 \exp(-E/RT) \quad , \tag{2.24}$$

where

D_0 = a constant

E = activation energy, cal/g mole

R = perfect gas low constant, 1.986 cal/g mole, K

T = absolute temperature, K.

According to Booth,[93] activation energy is essentially constant, but D_0 varies with the nature of the compact. By using this theory, he developed a method for computing fission gas release based on the diffusion equation for a spherical crystallite of radius a. He expressed his results in terms of F, the fraction of total gas produced which has diffused out of the sphere in time t. From $t = 0$ to $t = 1/(\pi^2 D')$ (at which time $F = 0.57$), fractional release can be approximated by

$$F = 4\left(\frac{D't}{\pi}\right)^{1/2} - \left(\frac{3D't}{2}\right).$$ (2.25)

For $t > (\pi^2 D't)^{-1} + 1$, Booth shows

$$F = \frac{0.57}{\pi^2 D't} + 1 - \frac{1}{\pi^2 D't} + \frac{6}{\pi^4 D't}\exp(-\pi^2 D't) - \frac{6}{\pi^4 D't}.$$ (2.26)

The pseudo-diffusion constant $D' = D/a^2$ is experimentally determined at a given temperature, T_0. The constant at any other temperature T is given by

$$D'(T, \text{K}) = D_0'\exp\left[E/R\left(\frac{1}{T_0} - \frac{1}{T}\right)\right].$$ (2.27)

Booth recommended a value of 45 000 cal/g mole for E.

Later studies indicated that the fission gas release process is much more complicated than originally believed. It has been suggested that a coupled diffusion-trapping process is the rate controlling mechanism.[91] In such a model, fission gas is assumed to be trapped in lattice defects; its release is partially controlled by its rate of escape from these traps. More recent models have been of the bubble migration type, which have been used for predicting fuel swelling. Thus, the BUBL (Ref. 88) and GRASS (Ref. 90) codes can be used for predicting both fission gas release and swelling. However, a critical survey by Ronchi and Matzke[94] estimates that most of the fission gas is kept in the matrix by resolution and that atomic diffusion accounts for a large fraction of the release. If this view is correct, models like BUBL-1, which describe fission gas release solely in terms of bubble migration, would be inappropriate. Models like GRASS, which account for resolution and atomic diffusion, could appropriately describe the release process if calibrated against experimental data.

Because of the difficulty in calibrating models like GRASS to available experimental data, most designers prefer to use simpler empirical correlations or models in which coefficients have been fitted to various in-reactor studies. Some of the correlations proposed are:

1. Lewis'[95] simple temperature zone model. The gas release fraction, F, is given by

$$F = (0.005 \ V_1 + 0.1 \ V_2 + 0.6 \ V_3)\sqrt{t/3} + 0.95 \ V_4 \quad , \tag{2.28}$$

where V_1, V_2, V_3, V_4 are fractional volumes of fuel in the temperature regions $<1000°C$, 1000 to 1300°C, 1300 to 1600°C, and $>1600°C$, respectively, and t is irradiation time in years.

2. Hoffman and Coplin's[96] correlation between release fraction and average fuel temperature during maximum power.

3. Beyer and Hann[97] temperature zone model. In this correlation, the gas release fraction, F, is given by

$$F = 0.05 \ X_1 + 0.141 \ X_2 + 0.807 \ X_3 \quad , \tag{2.29}$$

where X_1, X_2, X_3 are fractional amounts of gas produced between 1200 to 1400°C, 1400 to 1700°C, and $>1700°C$, respectively.

4. A probabilistic approach developed by Weisman et al.,[98] which assumes that (a) proportion K_1 of the fission gas escapes without being trapped and (b) fraction K_2 of the trapped particles is released per unit time. The rate of gas release, r, per unit time is given by

$$r = p'\left\{ t - \frac{1 - K_1}{K_1 K_2}[1 - \exp(K_1 K_2)]\right\} \quad , \tag{2.30}$$

where

p' = gas production rate

t = time.

By assuming that reactor power operation can be described by a series of constant power steps, total gas release can be determined. The number of moles released, Δr_i, during the i'th time interval, Δt_i, is then given by

$$\Delta r_i = r_i - r_{i-1} = p_i'\left\{ \Delta t_i - \frac{1 - K_1}{K_1 K_2}[1 - \exp(-K_1 K_2 \Delta t_i)]\right\} + C_{i-1}[1 - \exp(-K_1 K_2 \Delta t_i)] \quad ,$$

$$\tag{2.31}$$

where C is gas concentration in fuel equal to $p't - r$. Since the total gas release from time zero is $\sum_i \Delta r_i$, the fraction of released gas produced is given by

$$F = \frac{\sum_i \Delta r_i}{\sum_i p_i \Delta t_i} \quad .$$

MacDonald and Wang[47] give constants K_1 and $K_1 K_2$ for 95% dense fuels as

$$K_1 = \exp(-12\,450/T + 1.84)$$
$$K_1 K_2 = 0.25 \exp(-21\,410/T) \quad,$$

where T is in °R and t is in hours.

5. Gruber's[99] parametric relationship based on the results of the FRAS bubble migration model. The model is designed for determining release behavior during an accident. It is assumed that initial gas concentration C_0 in atom/cm^3 is known. Gruber concludes that for a typical grain size of 4 μm,

$$dF/dt = a_1 \nabla T \exp[(-a_2/T) - (a_3 - a_4 T)F] \quad, \tag{2.32}$$

where

t = time, s

∇T = temperature gradient, K/cm

T = temperature, K

F = fractional release

$a_1 = 7.68 \times 10^{18}\, C_0^{-0.7654}$, cm/K s

$a_2 = 37\,800$, K

$a_3 = -1833.3 + 77.83 \ln C_0 - 0.8239\,(\ln C_0)^2$

$a_4 = 0.0227 - 4.894 \times 10^{-4} \ln C_0$, K^{-1}.

Once the total fission gas release[†] has been obtained, the internal pressure and deflection can be calculated. Account should be taken of the thermal expansion of fuel and cladding and the elastic expansion of the cladding. Thus, for a cylindrical rod in which the internal cladding diameter is greater than pellet diameter

$$V_c(1 + 3a_c \Delta T_c) - V_f(1 + 3a_f \Delta T_f) = \frac{N^* R T}{P_i} - \frac{\pi D^3 L}{4EC}\,(P_i - P_0) \quad, \tag{2.33}$$

where

V_c = cold cladding volume

ΔT_c = hot cladding-cold cladding temperature

V_f = cold fuel volume

a_c, a_f = cladding and fuel linear coefficient of thermal expansion, respectively

ΔT_f = hot fuel-cold fuel temperature

N^* = number of gram moles of gas that have escaped

D = hot inside diameter of cladding

R = gas constant, 73.8 lb in./g mole K

[†]Recent data from high burnup fuel rods indicate that the empirical models, which are based on low burnup tests, tend to underpredict the fission gas release fraction at high burnups. Meyer et al. [*Nucl. Safety*, 19, 699 (1978)] recommend that predictions based on such correlations be increased by use of a burnup-dependent enhancement factor whenever the burnup exceeds 20 000 MWd/tonne.

T = temperature of gap, K

C = cladding thickness

L = cladding length

P = external pressure

P_i = internal pressure

E = Young's modulus of elasticity.

The above does not include the effects of irradiation swelling of the fuel or of cladding creep, which must be considered when Zircaloy is used.

In view of the radiation embrittlement of Zircaloy and stainless steel, a limitation is generally placed on cladding strain during normal operation. At higher temperatures, where ductility is greater but strength is lower, cladding bursting under external depressurization must be evaluated.

In fuel plates, unacceptable deformation of the plates is a limit. Lustman,[100] who assumed that plate compartments deform into trapezoidal prisms, obtained the following expressions for maximum deformation y and internal pressure P_i

$$y = \frac{0.0138\,a^4}{Ec^3}\left(\frac{N^*RT}{V_0 - 2ya^2/3} - P_0\right) \tag{2.34}$$

$$P_i = (N^*RT)/(V_0 + 2ya^2/3) \quad , \tag{2.35}$$

where

V_0 = initial free volume in plate element (cold condition), in.^{3}

a = edge length of plate compartment square, in.

Other symbols have their previous meanings. Creep has again been omitted.

2-2.5 Thermal Resistance of Fuel and Fuel-Cladding Gap

With pressed and sintered pellets, there is usually an appreciable resistance to heat transfer between the pellet surface and cladding. Since bonding agents are not used in PWR design, interfacial resistance can be that of a gas-filled gap or UO_2 in actual contact with the cladding surface. It has often been assumed that the pellet remains nearly centered within the rod and that gap conductance can be obtained from the conductance of an annular gas gap. If this assumption is correct, then gap conductance h_g can be estimated from

$$h_g = \frac{k_g}{l + l_0} \quad , \tag{2.36}$$

where

k_g = thermal conductivity of gas in gap, Btu/h °F ft

l = gap at operating conditions

l_0 = constant specified to provide a conductance consistent with that computed at zero contact pressure.

Cohen et al.[101] investigated gap conductance for UO_2 stainless-steel-clad fuel elements under operating fluxes. The operating gap was obtained by correcting for the thermal expansion of cladding and fuel and by assuming the fuel remains centered within the cladding. Figure 2-11 shows that data can be fitted by an equation like Eq. (2.36). When determining the gap at operating conditions, an additional correction to cladding diameter should be made to account for fuel and cladding creep and for elastic deformation of the cladding due to the differential between external and internal pressure. For an infinitely long cylinder, subject to internal and external pressure, elastic deformation, ΔD_i, is given by

$$\Delta D_0 = \frac{D_i}{E(K^2 - 1)}\left\{P_i[(1 - \gamma) + (1 + \gamma)K^2] - 2K^2 P_0\right\} , \qquad (2.37)$$

where

D_i = internal diameter, in.

D_0 = external diameter, in.

P_i = internal pressure, lb/in.^{2}

P_0 = external pressure, lb/in.^{2}

E = Young's modulus for the cladding

$K = D_0/D_i$

γ = Poisson's ratio for the cladding.

When low molecular weight gases (He, H_2) are present in the small-sized gaps usually encountered, a correction should be made to account for temperature discontinuity at the gas-solid interface. Knudsen[102] postulated that the energy interchange between gas molecules and a solid surface may be incomplete. He defined the accommodation coefficient α as the ratio of the temperature differential of the incident and reflected gas molecules to the temperature difference between the solid and incident molecules. Dean[103] proposed that Eq. (2.36) be modified to

$$h_g = \frac{k_g}{f'(l + l_0)} , \qquad (2.38)$$

where f' is the factor by which bulk thermal conductivity should be divided to obtain an effective conductivity that allows for temperature discontinuity. Dean shows f' to be given by

$$f' = 1 + \left(\frac{\alpha_1' + \alpha_2' - \alpha_1'\alpha_2'}{\alpha_1'\alpha_2'}\right)\left(\frac{4}{c_p + c_V}\right)\left(\frac{k_g}{\mu}\right)N_{kn} , \qquad (2.39)$$

where

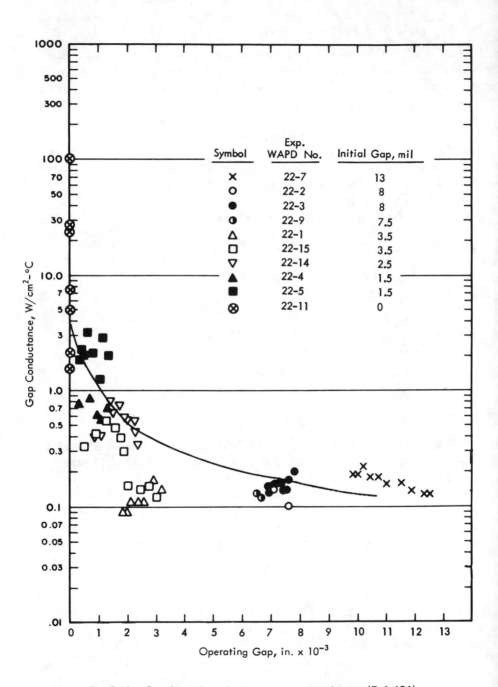

Fig. 2-11.   Gap thermal conductance versus operating gap (Ref. 101).

$\alpha_1', \alpha_2' =$ accommodation coefficients for fuel and cladding surfaces

$c_p, c_V =$ specific heats at constant pressure and constant volume, respectively

$\mu =$ gas viscosity, lb/ft h

$N_{kn} =$ Knudsen number, Ω/l

$\Omega =$ mfp of gas molecules, ft.

Lanning and Hann[104] suggest the accommodation coefficient can be taken as a property of fill gas only. Based on the data of Ullman et al.,[105] they suggest

$$\alpha_{He}' = 0.425 - 2.3 \times 10^{-4} \, T_s \qquad (2.40)$$

$$\alpha_{Xe}' = 0.749 - 2.5 \times 10^{-4} \, T_s \quad , \qquad (2.41)$$

where T_s is surface temperature, K. Accommodation coefficients for other gases are obtained by linear interpolation between α_{He}' and α_{Xe}' based on molecular weight. They obtain the accommodation coefficient for a gas mixture based on the mean molecular weight of the mixture. Godesor et al.[85] suggest the accommodation coefficients vary somewhat with the type of surface.

MacDonald and Thompson[50] propose that f' be calculated from

$$f' = 1 + C(k\sqrt{T_g}/Pl) \quad , \qquad (2.42)$$

where

$T_g =$ gas temperature, K

$P =$ pressure, N/m^2

$C =$ empirical constant.

For typical UO_2-Zr surfaces, they indicated that

$$C = 0.21 \frac{N(K)^{1/2}}{w} = 8.5 \times 10^{-4} \frac{\text{psi-ft}^2 \, \text{h} \, ^\circ F^{1/2}}{Btu} \quad .$$

Gas thermal conductivity can be determined from the data of Von Ubisch et al.,[106] who examined the thermal conductivity of binary and ternary mixtures of xenon, krypton, and helium. They concluded that a mixture of xenon and krypton, which approximates fission gas (15.3% Kr, 84.7% Xe), had a thermal conductivity, k_g, represented by

$$k_g = k_0(T/302)^s \quad , \qquad (2.43)$$

where

$T =$ gas temperature, K

$k_0 = 147$

$s = 0.86$

$k_g =$ gas thermal conductivity, 10^{-7} cal/cm^2 K s.

The thermal conductivity of helium could be represented by the same equation with $k_0 = 3670$ and $s = 0.72$. To estimate the thermal conductivity of mixtures of fission gas and helium, they suggest

$$k_m = (k_{He})^x (k_{FG})^{1-x} \quad , \tag{2.44}$$

where

k_m = thermal conductivity of mixture

k_{He} = thermal conductivity of helium

k_{FG} = thermal conductivity of fission gas

x = mole fraction helium.

Lanning and Hann[104] suggest a more theoretical approach, based on the Chapman-Enskog theory, for calculating mixture conductivities.

Providing the gap thickness is large with respect to the mean-free-path of the gas molecules, thermal conductivity is nearly independent of pressure.

When the pellet expands sufficiently to be in direct contact with the cladding, an improved gap conductance results. Several theoretical explanations of the phenomenon of contact conductance have been proposed and, in general, they consider a model in which the solid surfaces are in actual contact at only a few discrete points (Fig. 2-12). Since thermal conductivity of the solids on contact is generally much greater than the thermal conductivity of gas filling the interstices, heat flow tends to channel through the points of contact. When contact pressure is increased, peaks in contact are deformed and the contact points increase in size and number. The number and size of the contact points also depend on the surface finishes. At low contact pressures, the primary mode of heat transfer can be by conduction across the gas gaps. The effective distance between the surfaces depends on the size and shape of surface irregularities.

Various analytical approaches have been proposed for computing contact conductance. Cetinkale and Fishenden[107] presented an equation for the contact conductance of two metal surfaces based on a calculation of the isotherms at the contact points of the two solids. A somewhat different approach was taken by

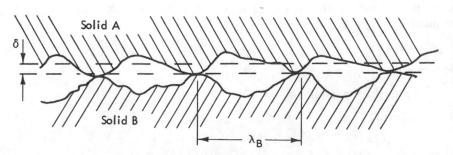

Fig. 2-12.   Contact conductance model.

Fenech and Rohsenow,[108] who found an approximate analytic solution for heat conduction at a contact point. The results obtained from these approaches have not been widely used by designers.

Few experimental measurements of direct use to the fuel rod designer are available. Wheeler[109] measured the contact conductance of seven pairs of materials, including cladding materials and UO_2. His results for experiments in a vacuum are plotted in Fig. 2-13 as a function of the ratio of contact pressure to yield strength. Rapier et al.[110] also examined the behavior of UO_2-stainless-steel interfaces.

Cohen's work[101] on irradiated specimens included measurements on a number of specimens with cladding-fuel contact. The contact conductance obtained for UO_2-stainless steel ranged from 4000 to 250 000 Btu/(h ft² °F) on the first startup. Contact resistance was evaluated by obtaining the total resistance and subtracting the resistances of the fuel and cladding. An apparent decrease in thermal conductance after the first startup may have been due to a decrease in fuel conductivity through irradiation or cracking.

Dean[103] measured the contact resistance of the UO_2-Zircaloy-2 joint as a function surface roughness and contact pressure in an argon atmosphere. In Fig. 2-14, we see that at the low contact pressure of the experiment, the conductance varied linearly with contact pressure and, as expected, increased with decreased surface roughness. Dean's measurements can be conservatively correlated as

$$h_g = bP + \frac{k_g}{14.4 \times 10^{-6}} \quad , \qquad (2.45)$$

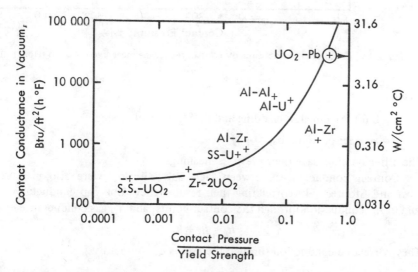

Fig. 2-13. Contact conductance measured by Wheeler. [From *Reactor Handbook,* Vol. IV, Interscience Publishers, Inc., New York (1964).]

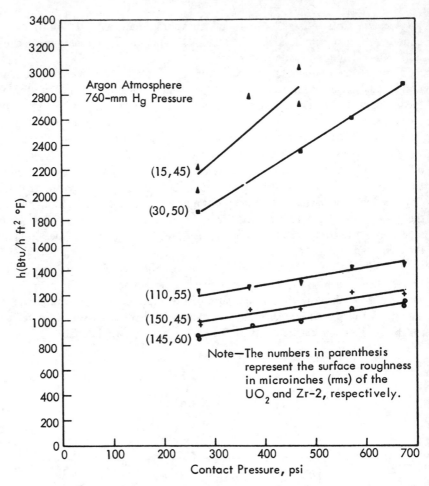

Fig. 2-14. Uranium dioxide-Zircaloy contact resistance measured by Dean (Ref. 103).

where

$b = 0.60$ for Zircaloy cladding and 0.48 for stainless steel

$P =$ contact pressure, psi.

The other symbols have their previous meanings.

Contact conductances between UO_2 and Zircaloy were also measured b:
Ross and Stoute.[111] In correlating their results, they took gap conductance as th
sum of h_s, conduction through the contact points, and h_f, gas conductance

$$h_g = h_s + h_f \quad . \tag{2.46}$$

They correlated conduction through the points of contact by

$$h_s = \frac{k_m P}{a_0 L_0^{1/2} H'} \quad , \tag{2.47}$$

where

k_m = harmonic mean of thermal conductivities k_1 and k_2 of contacting surfaces $[(2k_1k_2)/(k_1 + k_2)]$, Btu/h ft °F

P = contact pressure, psi

a_0 = empirical constant, 0.0905

L_0 = root mean square (rms) of surface projections, $[(g_1^2 + g_2^2)/2]^{1/2}$

H' = Meyer hardness of softer solid, psi.

Typical values of H' are 13×10^4 psi for stainless steel and 14.2×10^4 psi for Zircaloy. Gas conductance was given by

$$h_f = \frac{k_g}{(2.75 - 0.00176P)\,[(l_1 + l_2) + (g_1 + g_2)]} \,, \qquad (2.48)$$

where

k_g = thermal conductivity of gas in gap

$l_1 + l_2$ = temperature jump distances at surfaces 1 and 2.

The temperature jump distance is another means of accounting for temperature discontinuity between fluid in the gap and contact surfaces. Ross and Stoute estimated these quantities for helium, argon, and fission gases from their data. For temperatures between 150 and 300°C, $(l_1 + l_2)$ is $< 10 \times 10^{-4}$ cm for helium, 5×10^{-4} cm for argon, and $< 1 \times 10^{-4}$ cm for xenon. The distances are assumed proportional to the mfp of the gas molecules and, hence, inversely proportional to the temperature. Campbell and DesHaies[112] verified Ross and Stoute's temperature jump distance estimates. To use these data for gas mixtures, Veckerman and Harris[113] show that

$$l_m = \frac{\sum x_i l_i (M_i)^{1/2}}{\sum x_i/(M_i)^{1/2}} \,, \qquad (2.49)$$

where

l_m = temperature jump distance of the gas mixture

l_i = temperature jump distance of constituent gases

x_i = mole fraction

M_i = molecular weight.

The best general correlation of the available experimental data on contact conductance appears to be Shlykov's.[114] He assumes, as did previous investigators, that total conductance can be expressed as the sum of the fluid and contact points. Shlykov first defines a maximum relative gap thickness, X, where

$$X = \frac{2(g_1 + g_2)}{2l} = \frac{\delta_{max}}{2l} \tag{2.50}$$

and

g_1, g_2 = mean heights of rough projections of surfaces 1 and 2

l = temperature jump distance

δ_{max} = maximum gap thickness.

The temperature jump distance is given by

$$l = \left(\frac{2 - \alpha'}{\alpha'}\right) \frac{2}{Pr} \left(\frac{c_p/c_V}{c_P/c_V + 1}\right) \Lambda \quad, \tag{2.51}$$

where

α' = accommodation coefficient

Pr = Prandtl number

Λ = mfp of gas molecules

c_P = specific heat at constant pressure

c_V = specific heat at constant volume.

Shlykov uses the work of Shvets and Dyban,[115] who were able to relate X to Y, the ratio of maximum to effective gas gap, by

$$Y = \frac{10}{3} + \frac{10}{X} + \frac{4}{X^2} - 4\left(\frac{1}{X^3} + \frac{3}{X^2} + \frac{2}{X}\right) \ln(1 + X) \quad, \tag{2.52}$$

where

$Y = \delta_{max}/\delta_e$

δ_e = effective (equivalent) gap thickness to be used in calculating the heat transferred across the gas gap.

The Shvets and Dyban correlation is compared to some of the available data in Fig. 2-15. Using these data, the conductance of gas gap h_{gas} can be calculated by

$$h_{gas} = k_g/\delta_e \quad, \tag{2.53}$$

where k_g and δ_e have their previous meanings.

To obtain the heat transferred through the contact points, Shlykov assumes the true contact area is a function of the ratio of the surface projections $(g_1 + g_2)$ to the mean radius of contact spot a and the ratio of contact pressure P to the yield strength of the softer material, σ. Total gap conductance then becomes

$$h_g = k_g/\delta_e + f\left(P/\sigma, \frac{a}{g_1 + g_2}\right)\frac{\overline{k}}{d} \quad, \tag{2.54}$$

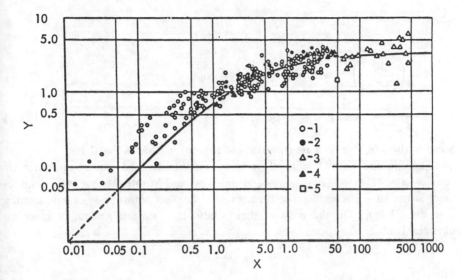

Fig. 2-15.   Test data on equivalent gap height in relative coordinates. Notation:

1. helium, argon, neon

2. air, helium, hydrogen

3. air

4. air

5. air.

[From *Therm. Eng.,* **12,** 102 (1966).]

where

$$\overline{k} = (2k_1 k_2)/(k_1 + k_2)$$

k_1, k_2 = conductivity of metal surfaces.

Two dimensionless groups are then defined as

$$N_c = h_g \delta_e / k_g \tag{2.55}$$

$$N_b = \overline{k} \delta_e / a k_g \ . \tag{2.56}$$

Equation (2.54) is then rearranged to yield

$$\frac{N_c - 1}{N_b} = f\left(\frac{P}{3}\right)\left(\frac{a}{g_1 + g_2}\right) . \tag{2.57}$$

By assuming a is 40 and correlating the available data, Shlykov obtains

$$\frac{N_c - 1}{N_b} = 0.32\left(\frac{PK}{3}\right)^{0.86} , \tag{2.58}$$

where the constant

$K = 1$, when $g_1 + g_2 \geqslant 30\mu$

$$K = \left(\frac{30}{g_1 + g_2}\right)^{1/3}, \text{ when } 10\mu \lesssim g_1 + g_2 \lesssim 30\mu \qquad (2.59)$$

$$K = \frac{15}{g_1 + g_2}, \text{ when } g_1 + g_2 \lesssim 10\mu \quad . \qquad (2.60)$$

Some of the available data are compared to this correlation in Fig. 2-16.

Lanning and Hann[104] critically review available methods for computation of contact conductance. They assume, as did Ross and Stoute,[111] that total conductance is the sum of conductance through solid contact points and gas conductance [see Eq. (2.46)]. On the basis of their study, they suggest gas conductance be obtained from

$$h_f = k_g/(\Delta X_{eff}) \quad , \qquad (2.61)$$

where

$\Delta X_{eff} = K_1[C(g_1 + g_2) + (l_1 + l_2)] + K_2$

$K_1 = 1.8$

$K_2 = -0.00012$

$C = 2 \exp(-0.00125\, P)$

$P = $ contact pressure, kg/cm^2

$g_1, g_2 = $ mean surface roughness (projection height) of fuel and cladding, respectively, cm

$l_1, l_2 = $ temperature jump distance, cm.

By comparing the available models (Dean, Ross and Stoute, Rapier, Mikic-Todreas,[116] Shlykov, Fenech-Rohsenow, Cetinkale and Fishenden) for solid conductance with seven sets of typical UO$_2$-metal data, they concluded that none of the models was really adequate. However, they found that when multiplied by $\frac{1}{4}$, the Mikic-Todreas[116] model was best and provided a generally conservative prediction of solid conductance data. Therefore, they recommended

$$h_s = 0.578\, k_m \left[\frac{g_1}{(g_1^2 + g_2^2)^{1/2}\lambda_c}\right](P/H)^n \quad , \qquad (2.62)$$

where

$k_m = $ harmonic mean of thermal conductivities of contacting surfaces

$\lambda_c = $ distance between peaks on cladding surface (Fig. 2-12)

$H = $ Meyer hardness of softer material, force/area

$P = $ contact pressure, force/area.

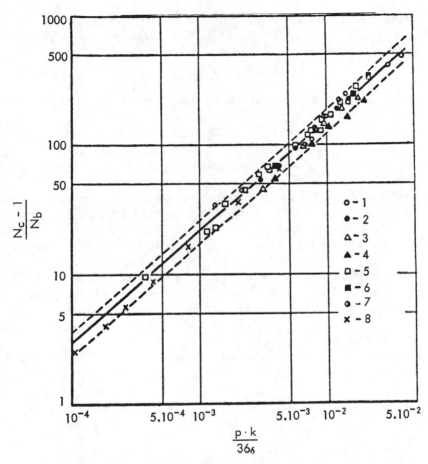

Fig. 2-16. Comparison of experimental data with Shlykov's contact conductance model. Notation:

1 through 4—steel

5 and 6—steel 30 CrMnSiN, different machining methods

7—Duralumin DT16

8—brass.

[From *Therm. Eng.*, **12**, 102 (Apr. 1965).]

At low contact pressures, $n = 0.5$, reflecting the idea that a fixed number of higher peaks per unit area control the conductance. At high pressures, n approaches 1.0 as an increase in the number of contact spots begins to be proportional to the increase in pressure. Todreas and Jacobs[116] indicate that n approaches 1.0 for Zircaloy surfaces at pressures above 1000 psi.

Contact pressure between the pellet and cladding can be estimated by computing their dimensions while ignoring any elastic deformation of the cladding.

The internal pressure, P_i, required to produce an elastic deformation that would produce a zero gap is given by Eq. (2.37). Since the cladding and pellet are in true contact over only a small fraction of the total area, we may say

$$P_i = P_{gas} + P_{contact} \quad .$$

Since UO_2 cracks so extensively at the heat generation levels at which the PWR operates, it has been suggested that it is unrealistic to assume that an annular gas gap can be retained even if calculations indicate there is a significant space between pellet and cladding. It is likely that cracked pellet pieces are lying against the cladding so gap conductance would then be computed from one of the contact-conductance equations, while assuming a zero contact pressure.

The data of Calza-Bini et al.[117] support this point of view. They observed that on the first rise to power, Eq. (2.36) provided reasonable estimates of the measured gap conductance. However, on subsequent power cycles, the measured gap conductance increased significantly. Stehle et al.[70] report that Brzoka and Wunderlich collected results showing somewhat similar behavior. They concluded that ~40% of the initial gap was eliminated during the first rise to power. After 500 h of irradiation, they concluded that another ~20% of the gap was closed. These estimates apparently made no allowance for a decrease in UO_2 thermal conductivity due to cracking. It is likely that such an allowance would lead to estimates of nearly complete gap closure after 500 h of operation.

Lanning et al.[118] avoided the assumption of an annular gap and simply correlated in-reactor data from 16 sources. For helium-filled rods, they suggest

$$h_g = 0.0321 + \frac{0.1167\sqrt{T_{gap}}}{2L'/D} - \frac{2.36}{2L'/D} \quad . \tag{2.63}$$

For krypton- and xenon-filled rods,

$$h_g = 0.1129 + \frac{0.1401}{2L'/D} + \frac{0.0069q'}{D} \quad , \tag{2.64}$$

where

T_{gap} = gap temperature, K

q' = linear heat rating, W/cm

L' = gap thickness

D = tube diameter

and h_{gap} is given in W/cm^2 °C.

MacDonald and Weisman[45] proposed a fairly complete model for cracked pellet behavior. They conclude that after a short period of initial operation at power, the pellet cracks and segments move outward, lying against the cladding. This causes an increase in gap conductance, but a decrease in fuel conductivity. They indicate that when the uncracked pellet diameter is less than the cladding diameter, gap conductance can be represented by

$$h_g = \frac{fk_g}{L_0'} + (1 - f)\frac{k_g}{L' + L_0'} \quad , \tag{2.65}$$

where

f = fraction of the pellet surface in contact with the cladding at zero contact pressure

L_0' = root mean square of fuel and cladding surface projections

L' = gap thickness neglecting pellet cracking or eccentricity.

Based on Kjaerheim and Rolstaad's suggestion,[119] the portion of fuel in contact with the cladding is obtained from

$$f = C_1 + (1 - C_1)C_2^{(L'C_3/D)} \quad , \tag{2.66}$$

where

C_1 = fraction of cracked fuel pellet assumed to be in contact with the cladding, which is independent of gap size

C_2, C_3 = constants representing the rate at which contact area fraction decreases with increasing gap

D = hot pellet diameter.

For the data examined, MacDonald and Weisman concluded $C_1 = 0.3$, $C_2 = 0.2$, and $C_3 = 50$. Once the gap closed ($L' < 0$), gap conductance was represented by a simplified Ross and Stoute[111] model.

By assuming that most of the annular gap is replaced by crack area in the pellet, MacDonald and Weisman[45] derived an effective thermal conductivity, k_e, for the cracked pellet. They obtained

$$k_e = k_{UO_2} \left\{ \frac{1}{\dfrac{2L' - A}{D[B(k_g/k_{UO_2}) + C]} + 1} \right\} \quad , \tag{2.67}$$

where

k_{UO_2} = thermal conductivity of uncracked UO_2

k_g = thermal conductivity of gas

D = diameter of fuel pellet.

With L' as gap thickness in mils and D expressed in mils, they found that the constants were

$$A = 2.5 \quad ,$$

$$B = 0.077 \quad ,$$

$$C = 0.015 \quad .$$

For $2L' \leqslant 2.5$, $k_e \simeq k_{UO_2}$.

MacDonald and Weisman[45] found that the centerline fuel temperatures, calculated using their cracked pellet model, compared very well with centerline temperature measurements in several instrumented fuel rod irradiations.

2-2.6 Mechanical Behavior of Cladding

Calculating fuel-cladding interactions depends on an accurate description of both cladding and fuel behavior. Fuel pellets expand thermally and swell (possibly densify) and, when subject to pressure by the cladding, creep. Cladding expands thermally, deforms elastically, and creeps inward or outward depending on the direction of pressure forces.

The hexagonal close-packed lattice of α-Zircaloy exhibits a preferred crystallographic orientation that confers a pronounced anisotropy in some mechanical properties. The thermal expansion of α-Zircaloy is different in the axial and radial directions. Based on the data of Mehan and Weisinger[120] for Zircaloy-2, for the radial direction Ref. 50 suggests

$$\Delta L/L = 8.2076 \times 10^{-4} - 7.856 \times 10^{-6}T + 1.204285 \times 10^{-8}T^2 - 6.1409 \times 10^{-12}T^3 \quad ,$$

$$(2.68)$$

and, for the axial direction,

$$\Delta L/L = -7.08257 \times 10^{-3} + 5.223606 \times 10^{-6}T + 1.204285 \times 10^{-9}T^2 \quad , \quad (2.69)$$

where $\Delta L/L$ is change in length per unit length and T is in K.

The Zircaloy alpha-beta phase transition starts at $\sim$1040 K and ends at $\sim$1270 K. Above 1270 K, both axial and radial expansion of beta-Zircaloy can be taken as $\sim$9.7 $\times10^{-6}$ in./in. K.

At low loads, Zircaloy strain, ϵ, is proportional to stress, σ, divided by E, Young's modulus. MacDonald and Thompson[50] express E for alpha-phase Zircaloy as

$$E = 1.148 \times 10^{11} - 5.99 \times 10^7 T \quad , \quad (2.70)$$

where T is in K and E is in N/m^2. In the beta region,

$$E = 1.05 \times 10^{11} - 4.7245 \times 10^7 T \quad . \quad (2.71)$$

Plastic deformation of Zircaloy can be obtained from

$$\epsilon = (\sigma/K)^{1/n} \quad , \quad (2.72)$$

where K is a strength coefficient and n is a dimensionless strain-hardening coefficient. Values for these quantities from Busby's[121] data are

$$n = -1.3498 \times 10^{-2} + 5.152 \times 10^{-4}T - 5.5947 \times 10^{-7}T^2 \quad , \tag{2.73}$$

$$K = 1.398304 \times 10^{19} - 1.219965 \times 10^6 T - 96.6206T^2 \quad , \quad N/m^2 \quad . \tag{2.74}$$

Temperature T is in K. Transition from the elastic to plastic region occurs when plastic strain is greater than elastic strain.

Neutron irradiation has a substantial effect on the strength and ductility of Zircaloy. Irradiation leads to increases in the yield strength and decreases in ductility. Because of irradiation-induced ductility reduction and further possible reductions due to hydride formation, the maximum permissible strain must be set at a low level. Some irradiation experience[122] indicates that the maximum strain should not exceed ~1%, since some test rods have failed at strains of this magnitude.

It was previously observed that fast neutron irradiation also affects Zircaloy creep. These effects have been investigated by Ibrahim[123] and others. These data have been correlated by[124]

$$\dot{e} = K\phi[\sigma + B \exp(C\sigma)] \exp(-10\,000/RT)t^{-1/2} \quad , \tag{2.75}$$

where

$\dot{e}$ = transverse creep rate, m/m s

$K = 5.129 \times 10^{-29}$

$C = 4.967 \times 10^{-8}$

T = temperature, K

ϕ = fast neutron flux, n/m^2 s (E > 1.0 MeV)

σ = transverse stress, N/m^2

$B = 7.252 \times 10^2$

$R = 1.987$ cal/mol K

t = time, s.

Although irradiation-induced swelling is primarily a problem of fast reactors, exposure to fast neutron flux does result in changes in the axial length of fuel tubes. Hagrman[125] correlated available irradiation growth data by an equation of the form

$$\Delta L/L = A \left[\exp(T_0/T)\right] (\phi t)^{1/2}(1 - 3f_3)(1 + 2C_w) \quad , \tag{2.76}$$

where

$\Delta L/L$ = fractional change in length

$A = 1.407 \times 10^{-16}$ (neutrons/m^2)$^{-1/2}$

$T_0 = 240.8$ K

T = cladding temperature, K

ϕ = fast neutron flux, n/m^2 s ($E > 1.0$ MeV)

t = time, s

C_w = degree of cold work (fraction of cross-sectional area reduction)

f_3 = texture factor (0.05 is a typical value).

2-3 STEADY-STATE CONDUCTION UNDER IDEALIZED CONDITIONS

2-3.1 Heat Generation Within a Fuel Plate

2-3.1.1 Uniform Heat Generation

Let us consider the temperature distribution within the fuel of a thin fuel plate. If we ignore axial conduction, a one-dimensional treatment can be used. From a heat balance across a differential element (see Fig. 2-17), we have

$$\left(k + \frac{dk}{dx}dx\right)\left(\frac{dT}{dx} + \frac{d^2T}{dx^2}dx\right) - k\frac{dT}{dx} + q'''dx = 0 \quad , \tag{2.77}$$

where x is the distance from the fuel element centerline, q''' is the volumetric heat generation rate, and other symbols have their previous meanings.

After eliminating the second-order term, we simplify to

$$\frac{d}{dx}\left(k\frac{dT}{dx}\right) + q''' = 0 \quad . \tag{2.78}$$

By integrating on the assumption that q''' is constant and by imposing the boundary condition that dT/dx is zero at the centerline, we obtain

$$k\frac{dT}{dx} = -xq''' \quad , \tag{2.79}$$

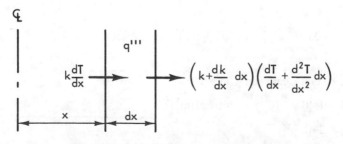

Fig. 2-17.   Heat conduction in a fuel plate.

and finally

$$\int_{T_c}^{T_x} kdT = -\frac{x^2}{2} q''' \quad , \tag{2.80}$$

where T_c is the center temperature. By substituting an appropriate expression for thermal conductivity, the temperature at any point can be determined. The assumption of negligible axial conduction can be justified for metallic fuel plates because of their very small thickness. For a UO_2 plate, the very low thermal conductivity of UO_2 provides the justification. The effect of axial conduction has been considered in detail by Fagan and Mingle,[126] who provide expressions for surface heat flux. They indicate that for most power reactor systems, the minimum heat flux will be in error by <0.5% if axial conduction is neglected.

2-3.1.2 Effect of Flux Depression

In a uniform lattice, diffusion theory indicates thermal neutron flux ϕ varies across the fuel plate in accordance with

$$\phi = \phi_0 \cosh(\kappa x) \quad , \tag{2.81}$$

where ϕ_0 is the flux at the centerline and κ is the reciprocal of the thermal diffusion length. Since q''' is proportional to ϕ in a thermal reactor,

$$q''' = q_0''' \cosh(\kappa x) \quad , \tag{2.82}$$

where q_0''' is the volumetric heat generation rate at the centerline. In actual practice, flux distribution probably will be computed by the escape probability method (see Sec. 1-2.5). Although a simple, closed form solution is not obtained, numerical results generally can be fitted quite well by equations like Eq. (2.81), provided κ is appropriately chosen.

If we proceed in the same manner as for a constant rate of heat generation, we obtain as the analog of Eq. (2.78)

$$\frac{d}{dx}\left(k \frac{dT}{dx}\right) + q_0''' \cosh(\kappa x) = 0 \quad . \tag{2.83}$$

Integration twice with $dT/dx = 0$ at the centerline yields

$$\int_{T_s}^{T_x} kdT = \frac{q_0'''}{\kappa^2} [\cosh(\kappa a) - \cosh(\kappa x)] \quad , \tag{2.84}$$

where a is the half thickness of the plate. We can also express our result in terms of surface heat flux per unit area q'' by noting that

$$q'' = \frac{q_0'''}{\kappa} \sinh(\kappa a) \quad , \tag{2.85}$$

hence,

$$\int_{T_s}^{T_x} k dt = q'' \left[\frac{\cosh(\kappa a) - \cosh(\kappa x)}{\kappa \sinh(\kappa a)} \right] , \tag{2.86}$$

and

$$\int_{T_s}^{T_c} k dt = q' \left[\frac{\cosh(\kappa a) - 1}{2a\kappa \sinh(\kappa a)} \right] , \tag{2.87}$$

where T_c is the center temperature.

2-3.2 Heat Generation Within a Fuel Rod

2-3.2.1 Constant Heat Generation Across a Rod

From a heat balance across a differential annular ring (Fig. 2-18), we have

$$\theta \left(k + \frac{dk}{dr} dr \right)(r + dr)\left(\frac{dT}{dr} + \frac{d^2T}{dr^2} dr \right) - \theta kr \frac{dT}{dr} + q''' r\theta dr = 0 . \tag{2.88}$$

By rearranging and omitting higher order terms, we obtain

$$\frac{1}{r} \frac{d}{dr}\left(kr \frac{dT}{dr} \right) + q''' = 0 . \tag{2.89}$$

This can be integrated to yield

$$kr \frac{dT}{dr} = -\int_{r_c}^{r} rq''' dr = -\frac{q'''}{2}(r^2 - r_c^2) . \tag{2.90}$$

$$\int_{T_r}^{T_c} k dT = \int_{r}^{r_c} \left[-\frac{q'''}{2}\left(r - \frac{r_c^2}{r} \right) \right] dr = \frac{q'''}{4} [(r^2 - r_c^2) - 2r_c^2(\ln r/r_c)] , \tag{2.91}$$

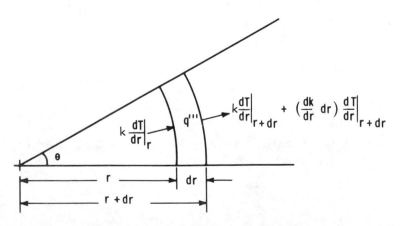

Fig. 2-18.   Heat conduction in a fuel rod. [From *Nucl. Eng. Des.*, **6**, 301 (1967).]

where T_c is the center temperature at central radius r_c and T_r is the temperature at radius r.

When we are dealing with solid cylinders (no central hole), we can simplify our last result to

$$\int_{T_s}^{T_c} kdT = \frac{q''' R_0^2}{4} \quad , \tag{2.92}$$

where R_o is outer radius and T_s is surface temperature.

We can also express the volumetric heat generation rate in terms of the rate of heat generation per unit length of fuel rod q' as

$$q' = \pi R_o^2 q''' \quad , \tag{2.93}$$

then

$$q' = 4\pi \int_{T_s}^{T_c} kdT \quad . \tag{2.94}$$

For the approximation of a constant value of k, we obtain

$$(T_c - T_s) = \frac{q'}{4\pi k_{av}} \quad . \tag{2.95}$$

From Eqs. (2.94) and (2.95), we find that the temperature difference across a fuel pellet is independent of the radius of the fuel pellet. Specifying a given surface temperature and a given rate of power generation per unit length of fuel fixes the central fuel temperature.

2-3.2.2 Concept of $\int kdT$

Equation (2.94) states that q', the rate of heat generation per unit length of a cylindrical fuel rod, is directly proportional to the thermal conductivity integral,

$$\int_{T_s}^{T_c} kdT \quad .$$

Thus, if values of the integral are available, the power level necessary to produce a given center temperature is readily specified.[127] This is convenient for the designer since it avoids the necessity to perform the integration of thermal conductivity values with respect to temperature. It is also a convenient way to report experimental data from in-pile irradiations. Under those conditions, power generation, center, and surface temperature can be known. Hence, $\int_{T_s}^{T_c} kdT$ is known, but pointwise values of k are not available. Data are usually given in terms of $\int_{0}^{T_c} kdT$, where

$$\int_{T_s}^{T_c} kdT = \int_{0}^{T_c} kdT - \int_{0}^{T_s} kdT \quad . \tag{2.96}$$

Several curves of the integral as a function of its upper limit are shown in Fig. 2-19.

While the concept is of particular value for UO_2 fuel rods, it is also useful in describing UO_2 fuel plate behavior. As seen from Eq. (2.80), a knowledge of the integral and plate thickness is sufficient to establish the volumetric power generation required for a given temperature. Indeed, it is possible to relate $\int kdT$ to the heat output for any geometric shape. Thus, for constant heat generation

$$\int_{T_s}^{T_c} kdT = Cq' \ ,$$
(2.97)

where C depends on fuel geometry. Specification of the maximum allowable value of the integral fixes the power capability of the fuel in any given geometry. As previously noted, a common design criterion is the avoidance of center melting. Since this is an observable point, a series of tests at varying power levels can be used to determine the $\int kdT$ at which this first occurs. Observations are not dependent on a precise knowledge of fuel conductivity or temperature distribution within the fuel. Values, determined by several investigators, for the $\int kdT$ to produce melting are shown in the upper right of Fig. 2-19. More recently, Lyons et al.[128] evaluated this integral in a series of short-time, in-pile tests. From metallographic examinations, they concluded that for 95% dense, sintered UO_2 pellets, the average value of the conductivity integral from 0°C to melting was 93.5 W/cm. Statistically determined 95% confidence limits were ±3.5 W/cm. Direct measurement of the integral based on a gas-bulb sensor yielded a value of ~90 W/cm. Vibratory compacted fuel was found to have a significantly lower integral even when the density correction was made.

Expressions for the thermal conductivity integral for sintered pellets have been provided by Lyons et al.[128] and MacDonald and Thompson,[50] among others. For 95% dense UO_2 with temperatures between 0 and 1650°C, MacDonald and Thompson suggest

$$\int_0^T kdT = 40.4 \ln (464 + T) + 0.027366 \exp(2.14 \times 10^{-3}T) - 248.02 \ .$$
(2.98)

For temperatures between 1650°C and melting,

$$\int_{1650}^T kdT = 0.02(T - 1650) + 0.027366 \exp(2.14 \times 10^{-3}T) - 0.93477 \ .$$
(2.99)

In the above, T is in °C and $\int kdT$ is in W/cm. This equation yields 95.4 W/cm as the conductivity integral from 0°C to melting.

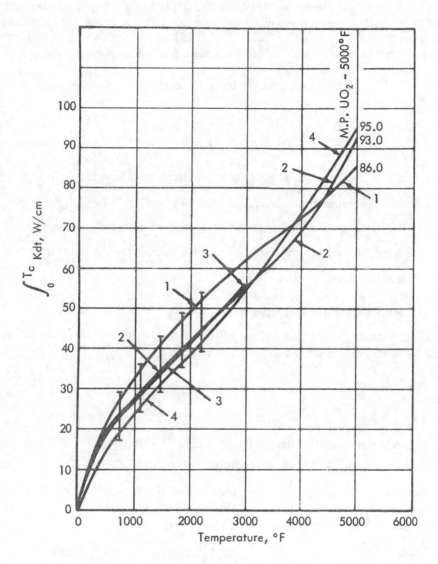

Fig. 2-19. Summary of $\int kdT$ for UO_2. Notation:

1. Weidenbaum, GEAP-3771-8, General Electric Company (1963)

2. Duncan, CVNA-142, Westinghouse Electric Corporation (1962)

3. Reference 39

4. Robertson et al., *J. Nucl. Mater.*, 7, *3*, 225 (1962).

[From McNelly, "Liquid Metal Fast Breeder Reactor Design Study," GEAP-4418, General Electric Company (1963).]

The effect of long-time irradiation on the conductivity integral is uncertain. While most earlier observers concluded there is significant reduction due to irradiation, some data dispute this.[44,129] Stora's previously cited explanation[44] that most apparent effects are simply due to thermal cracking and, thus, depend on available gap, seems most reasonable.

Soulhier[130] used the conductivity integral for correlation of fission gas release data. His results are presented as

$$\int_{400°C}^{T_C} kdT \text{(max) versus } F\int_{T_s}^{T_C} kdT \quad,$$

where F is the fractional gas release, T_C is center temperature, and T_s is surface temperature. The integral $\int_{400°C}^{T_C} kdT$ is used to normalize all experiments performed at a given central fuel temperature to the same surface temperature of 400°C. The product $F\int_{T_s}^{T_C} kdT$ is used to normalize the fission gas released based on the volumetric average heating rate.

2-3.2.3 Rod with Flux Depression

We previously showed that diffusion theory indicates thermal flux distribution across a cylindrical fuel rod in a uniform lattice given by

$$\phi = AI_0(\kappa r) \quad, \tag{2.100}$$

where

A = constant

I_0 = zero-order modified Bessel function of the first kind

κ = the reciprocal of the thermal diffusion length in the fuel.

When the escape probability method is used, which is more usually done, κ should be regarded as a constant selected to yield a closed-form approximation of the results. Since q''' is proportional to ϕ in a thermal reactor

$$q''' = A'I_0(\kappa r) \quad. \tag{2.101}$$

Now let q_0''' be the value of q''' at the center of the pellet. Since $I_0(0) = 1$, our previous expression becomes

$$q''' = q_0'''I_0(\kappa r) \quad. \tag{2.102}$$

The average value of q''' over the entire rod cross section is given by

$$\bar{q}''' = \frac{\int_0^a q'''(2\pi r)dr}{\int_0^a 2\pi r dr} = \frac{2}{a^2}\int_0^a q'''r dr \quad, \tag{2.103}$$

where a is the outer radius of the pellet.

By substituting our expression for local flux in Eq. (2.103), we obtain

$$\overline{q}''' = \frac{2q_0''' \, [I_1(\kappa a)]}{(\kappa a)} \quad . \tag{2.104}$$

Substituting the heat flux expression in our equation for the temperature gradient at any point [Eq. (2.89)] and integrating yield

$$\frac{q_0''' r [I_1(\kappa r)]}{(\kappa r)} + k \frac{dT}{dr} = 0 \quad . \tag{2.105}$$

On separating variables and integrating from r to a, we obtain

$$\frac{q_0'''}{\kappa^2} \left[I_0(\kappa a) - I_0(\kappa r) \right] = \int_{T_S}^{T} k dT \quad . \tag{2.106}$$

If we rewrite the above in terms of the average heat generation rate, we get

$$\frac{\dfrac{\overline{q}''' a^2}{4} \left\{ 2 \left[I_0(\kappa a) - I_0(\kappa r) \right] \right\}}{\kappa a \left[I_1(\kappa a) \right]} = \int_{T_S}^{T} k dT \quad . \tag{2.107}$$

At the center of the pellet, $r = 0$ and $T = T_C$. Under these conditions, Eq. (2.107) becomes

$$\left(\frac{\overline{q}''' a^2}{4} \right) \frac{2 \left[I_0(\kappa a) - 1 \right]}{\kappa a \left[I_1(\kappa a) \right]} = \int_{T_S}^{T_C} k dT \quad . \tag{2.108}$$

If we let

$$f = \frac{2 \left[I_0(\kappa a) - 1 \right]}{\kappa a \left[I_1(\kappa a) \right]} \quad , \tag{2.109}$$

and replace $\overline{q}'''$ in terms of q', the rate of heat generation per unit length of rod [Eq. (2.93)], we obtain

$$q' f = 4 \pi \int_{T_S}^{T_C} k dT \quad . \tag{2.110}$$

The term f, which is the "flux depression factor," becomes clear when we compare Eqs. (2.94) and (2.110). We see the thermal power associated with given fuel surface and center temperature is greater for nonuniform than for uniform heat generation; the ratio of the two heat ratings is f.

Approximate values of f can be obtained from diffusion theory if values of κ are available. For a UO_2 of 95% theoretical density, κ varies from 2 to 3 cm^{-1} for enrichment values of 2.5 to 6% ^{235}U. Figure 2-20 shows f as a function of enrichment and fuel diameter for pellets of 95% dense UO_2. For the enrichments used in PWRs, the value of f is not far from unity. However, this effect should not be

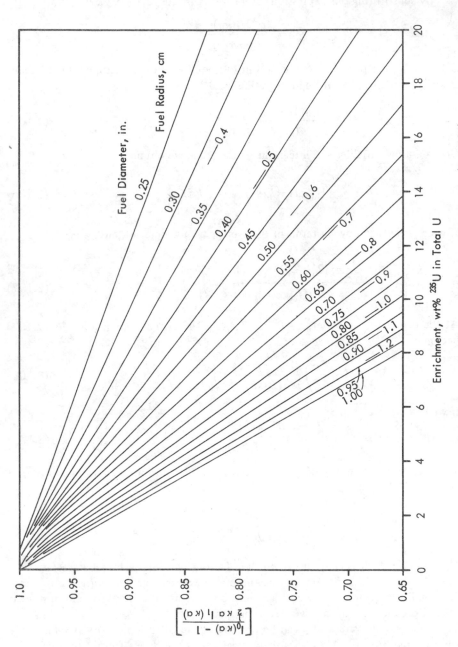

Fig. 2-20. Flux depression factors for UO_2 fuel rods of 95% theoretical density (Ref. 127).

ignored since it can mean an increase of several percentages in reactor power output. The fact that f is close to unity can be used to justify the simple approximation that thermal flux inside the fuel follows a parabolic distribution. If this is done, values of F (the ratio of surface-to-average flux in the rod) obtained from escape probability theory can be used directly to compute f without assigning a value to κ. Consequently, for a parabolic flux distribution

$$f = (3 - F)/2 \quad . \tag{2.111}$$

2-3.2.4 Heat Generation with a Power Tilt

In Sec. 1-3.2, it was indicated that flux distribution across a cluster of rods in a pressure tube of a D_2O or graphite-moderated reactor is approximated by the function $I_0(r)$. When a single rod is considered (Fig. 2-21), this appears as a flux tilt. A reasonable representation of the distribution has been found to be[131]

$$q'''(r,\theta) = \bar{q}'''(1 + \epsilon r/a \cos\theta) \quad , \tag{2.112}$$

where ϵ represents half the difference between the minimum and maximum heat generation rate divided by q''', the average volumetric heat generation rate

$$\epsilon = (q'''_{max} - q'''_{min})/2\bar{q}''' \quad . \tag{2.113}$$

Since we now consider both radial and azimuthal variation in temperature, we must use the general form of the steady-state conduction equation

$$k\nabla^2 T + q''' = 0 \quad . \tag{2.114}$$

In cylindrical coordinates with the present heat flux distribution, this becomes

$$\frac{1}{r}\frac{\partial}{\partial r}\left(r\frac{\partial T}{\partial r}\right) + \frac{1}{r^2}\frac{\partial^2 T}{\partial \theta^2} + \frac{\bar{q}'''}{k}\left(1 + \epsilon\frac{r}{a}\cos\theta\right) = 0 \tag{2.115}$$

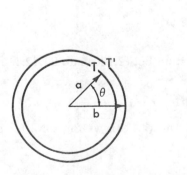
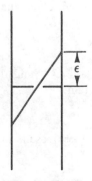

Fig. 2-21.   Fuel rod with flux tilt.

in the pellet region, if we assume a constant value for k.[a] Since we have no heat generation in the cladding region, we obtain

$$\frac{1}{r}\frac{\partial}{\partial r}\left(r\frac{\partial T'}{\partial r}\right) + \frac{1}{r^2}\left(\frac{\partial^2 T'}{\partial \theta^2}\right) = 0 \quad , \tag{2.116}$$

where T' represents cladding temperature.

At the boundary of the pellet and cladding, there is normally a gap to which we assign a conductance h_g. Since heat transferred across the pellet, gap, and cladding must be equal at $r = a$, we have the boundary conditions at $r = a$

$$h[T(a,\theta) - T'(a,\theta)] = k\left(\frac{\partial T}{\partial r}\right)\bigg|_{r=a} \tag{2.117}$$

$$k'\frac{\partial T'}{\partial r}\bigg|_{r=a} = k\frac{\partial T}{\partial r}\bigg|_{r=a} \quad , \tag{2.118}$$

where k and k' are the thermal conductivity of the fuel and cladding, respectively. If the heat transfer coefficient to the coolant is high, at $r = b$ we say

$$T'(b,\theta) \approx T_{\text{bulk water}} \quad . \tag{2.119}$$

Also note that at $r = 0$, T is not infinite. With these boundary conditions, Eqs. (2.115) and (2.116) can be integrated to yield

$$T = \frac{\bar{q}'''}{4k}(a^2 - r^2) + \frac{1}{2h}a\bar{q}''' + \frac{\bar{q}'''a^2}{2k'}\ln\frac{b}{a} + \psi(r)\cos\theta + T_b \quad , \tag{2.120}$$

and

$$T' = \frac{\bar{q}'''a^2}{2k}\ln\frac{b}{r} + \psi'(r)\cos\theta + T_b \quad . \tag{2.121}$$

We must now evaluate $\psi(r)$ and $\psi'(r)$ by substituting Eq. (2.120) into Eq. (2.115) and solving the resulting differential equation to get

$$\psi(r) = Ar + \frac{\alpha}{8}r^3 \quad , \tag{2.122}$$

where $\alpha = \epsilon\bar{q}'''/ak$. Similarly, if we substitute Eq. (2.120) into Eq. (2.116), we can solve to obtain

$$\psi'(r) = C(1/r - r/b^2) \quad . \tag{2.123}$$

Using these results in Eqs. (2.117) and (2.118), we can solve for constants A and C

---

[a] Andrews and Dixmier[132] provide a solution for the case of varying k.

$$A = \frac{\alpha a^2}{8} \left[\frac{1 + 3\left(\frac{k}{h_g a} + \frac{k}{k'}\frac{b^2 - a^2}{b^2 + a^2}\right)}{1 + \left(\frac{k}{h_g a} + \frac{k}{k'}\frac{b^2 - a^2}{b^2 + a^2}\right)} \right] \qquad (2.124)$$

and

$$C = \frac{a^4}{4}\frac{k}{k'} \left[\frac{b^2/(b^2 + a^2)}{1 + \left(\frac{k}{h_g a} + \frac{k}{k'}\frac{b^2 - a^2}{b^2 + a^2}\right)} \right] \, . \qquad (2.125)$$

If desired, the maximum fuel temperature at any radius can now be determined. This occurs at $\theta = 0$,

$$T_{max} = \frac{\bar{q}'''}{4k}(a^2 - r^2) + \frac{1}{2h_g}a\bar{q}''' + \frac{\bar{q}'''a^2}{2k'}\ln\frac{b}{a} + T_b + \frac{\epsilon\bar{q}'''}{8ak}\left(\frac{1 - 3\xi}{1 + 9\xi}a^2 r_{max} - r_m^3\right) \, , \qquad (2.126)$$

where

$$\xi = \frac{k}{h_g a} + \frac{k}{k'}\frac{b^2 - a^2}{b^2 + a^2} \, , \qquad (2.127)$$

and r_m, the radial location of the maximum fuel temperature, is given by

$$r_m = \frac{2a}{3\epsilon}\left\{\left[1 + 3/4\epsilon^2\left(\frac{1 + 3\xi}{1 + \xi}\right)\right]^{1/2} - 1\right\} \, .$$

We can also determine the maximum heat flux [Btu/(h ft^2)] that also occurs at $\theta = 0$,

$$q''_{max} = k'\left(\frac{2T'}{2r}\right)_{a=b} = \frac{\bar{q}'''a^2}{2b} + \frac{1}{2}\epsilon\bar{q}'''a^3[(1 + \xi)(b^2 + a^2)]^{-1} \, . \qquad (2.128)$$

Hence,

$$\frac{q''_{max}}{q''_{mean}} = 1 + \frac{ab\epsilon}{(1 + \xi)(b^2 + a^2)} \, . \qquad (2.129)$$

For $b = a + \Delta r'$, where $a \gg \Delta r'$, the last equation simplifies to

$$\frac{q''_{max}}{q''_{mean}} = 1 + \frac{\epsilon}{2\left[1 + \frac{k}{a}\left(\frac{1}{h_g} + \frac{\Delta r'}{k'}\right)\right]} \, . \qquad (2.130)$$

An example of what is obtained in a typical case is useful: Assume $\epsilon = 0.2$, $k' = 12$ Btu/h ft °F, $k = 1.6$ Btu/h ft °F, $a = 0.4$ in., $h_g = 1000$, $\Delta r' = 0.013$, then q_{max}/q_{mean} is ~ 1.09.

2-3.3 Annular Fuel Elements

For a hollow cylinder of outer radius a and inner radius b, diffusion theory calculations give the flux distribution as

$$\phi = C'\phi_0 \left[I_0(\kappa r) + \frac{I_1(\kappa b)}{K_1(\kappa b)} K_0(\kappa r) \right] , \tag{2.131}$$

where

K_0 = zero-order modified Bessel function of second kind

K_1 = first-order modified Bessel function of second kind

C' = constant evaluated so that $\phi = \phi_0$ at $r = a$.

For heat removal at the outer surface only, we write

$$\int_{T_s}^{T_c} k dT = \frac{1}{4r} q' \left\{ \frac{\kappa b \left[I_1(\kappa b) K_0(\kappa a) + I_0(\kappa a) K_1(\kappa b) \right] - 1}{\frac{1}{2} \kappa^2 ab \left[I_1(\kappa a) K_1(\kappa b) - I_1(\kappa b) K_1(\kappa a) \right]} \right\} . \tag{2.132}$$

When heat is removed at both surfaces, the situation is complicated if different boundary conditions hold at the two surfaces. When T_s is the same at both surfaces, we get

$$\int_{T_s}^{T_r} k dT$$

$$= \frac{1}{2} q' b \left(\frac{\left[I_0(\kappa b) - I_0(\kappa r) \right] + \left[K_0(\kappa b) - K_0(\kappa r) \right] + \kappa R^0 \ln \frac{r}{b} \left[I_1(\kappa r) - \lambda K_1(\kappa R^0) \right]}{\frac{1}{2} \kappa b \left\{ \frac{R^0}{b} \left[I_1(\kappa R^0) - \lambda K_1(\kappa R^0) \right] - \left[I_1(\kappa b) - K_1(\kappa b) \right] \right\}} \right) , \tag{2.133}$$

where

$\lambda = I_1(\kappa b) / K_1(\kappa b)$

R^0 = radius at which maximum fuel temperature occurs.

We obtain R^0 from

$$R^0_{\ln} \left(\frac{a}{b} \right) \left[I_1(\kappa R^0) - \lambda K_1(\kappa R^0) \right] = \frac{1}{\kappa} \left[I_0(\kappa a) - I_0(\kappa b) \right] + \frac{\lambda}{\kappa} \left[K_0(\kappa a) - K_0(\kappa b) \right] . \tag{2.134}$$

2-4 REALISTIC EVALUATION OF STEADY-STATE TEMPERATURES

Fuel element temperature calculations of the previous section did not consider the slowly varying changes in fuel element properties that take place during reactor exposure. After the initial rise to power, cracking of the fuel pellet changes fuel conductivity and gap conductance. Gap thickness decreases with time as the Zircaloy cladding creeps down toward the fuel pellet. In most long irradiations, fuel swelling eventually brings the pellet into contact with the cladding and further increases gap conductance. In addition, at high heat ratings, a central void gradually forms. Because of these gradual changes, fuel element temperatures are a function of the power-time history of the fuel rod. Realistic fuel temperature evaluations must include all these time-varying effects.

2-4.1 Calculational Procedures

In view of the complexity of the phenomena involved, satisfactory closed-form analytical solutions are not available[b]; numerical procedures are required. The fuel pellet is usually divided into a series of concentric annular rings. Using an assumed temperature distribution (or that from a previous time period), the response of each of these rings to the combined effects of thermal expansion, swelling, void migration, creep, and hot pressing during some short time period is determined. The interaction between pellet and cladding is then determined to establish gap conductance. Temperature distribution is then redetermined. If the new distribution is close to that originally chosen, the calculation proceeds to the next time interval. If agreement is poor, the behavior of the fuel rings is redetermined and a revised temperature distribution is obtained. The process continues at a given time step until convergence is obtained.

The simplest situation arises when the center temperature is below ~1450°C. Under these conditions, fuel restructuring does not occur, no central void is formed, and fuel creep and hot pressing are minimal. Thus, the change in fuel dimensions is often taken as the sum of that due to thermal expansion and swelling. An initial temperature distribution is assumed; thermal expansion and swelling of each ring is evaluated. Summation of the ring dimensions provides the new fuel pellet diameter. This is compared to cladding diameter and gap width is determined.

If a sufficiently large number of rings are chosen, properties are assumed constant across each annular ring. In one procedure,[134] it is assumed that heat flux q''_{av} and thermal conductivity k_{av} can be evaluated at the center of the annular ring and that the equation for the temperature drop across a slab can be used. Then the temperature drop across an annular ring, ΔT_r, is given by

$$\Delta T_r = \frac{q''_{av}(\Delta r)}{k_{av}} \ ,$$

(2.135)

where r is thickness of the ring.

When fewer rings are chosen, more accurate results are obtained by modifying Eq. (2.91) to

---

[b]Lamkin and Brehm[133] provide an exact closed-form solution for fuel temperatures and stresses, providing simplified creep and property formulation are accepted.

$$\int_0^{T_{r_1}} k\,dT = \int_0^{T_{r_2}} k\,dT + \left\{ \frac{q'''}{4} \left[(r_2^2 - r_1^2) - 2r_1^2(\ln r_2/r_1) \right] \right\} , \qquad (2.136)$$

where T_{r_1} and T_{r_2} equal temperatures at the inner and outer radius of the ring. If the temperature of $r_2(T_{r_2})$ is known, $\int_0^{T_{r_2}} k\,dT$ is known for a given porosity. Then $\int_0^{T_{r_1}} k\,dT$ can be obtained and T_{r_1} is known.

Both of the above schemes are iterative. A surface temperature of the pellet must be assumed. The calculation then proceeds inward. The final temperature distribution is then used to compute a revised estimate of fuel dimensions and gap conductance, h_g. With a new h_g, a revised pellet surface temperature can be computed.

In thermal reactors, irradiation swelling of the cladding is not significant. Thus, cladding strain is the sum of thermal expansion, elastic, and creep strain. Circumferential cladding strain, ϵ_θ, is given by

$$\epsilon_\theta = (1/E)[\sigma_\theta - \gamma(\sigma_r + \sigma_z)] + \alpha_c(\Delta T) + \epsilon_c , \qquad (2.137)$$

where

$\sigma_r, \sigma_\theta, \sigma_z$ = radial, tangential, and axial stresses, respectively

E = Young's modulus

α_c = coefficient of thermal expansion

ϵ_c = creep strain

γ = Poisson's ratio

ΔT = increase in cladding temperature.

Given a cladding with outer radius b and inner radius a, the stresses are obtained from

$$\sigma_r = \frac{a^2}{b^2 - a^2} (P_i - P_o)(1 - b^2/r^2) - P_o \qquad (2.138)$$

$$\sigma_\theta = \frac{a^2}{b^2 - a^2} (1 + b^2/r^2)(P_i - P_o) - P_o \qquad (2.139)$$

$$\sigma_z = \frac{\lambda}{\lambda + G} \left(\frac{1}{b^2 - a^2} \right)(a^2 P_i - b^2 P_o) , \qquad (2.140)$$

where

P_i, P_o = inner and outer pressures, respectively

$$G = \frac{E}{2(1 + \gamma)}$$

r = radius at a given point

$$\lambda = \frac{2G}{1 - 2\gamma}$$

E = Young's modulus.

Since a and b do not differ greatly, stresses at an average value of r can be used to estimate an approximate value of ϵ_r.

Equation (2.140) ignores interaction between the pellet stack and cladding. It is often concluded that friction between fuel and cladding prevents axial strain of the cladding.

The value of P_o is determined by external system pressure. When the fuel pellet is not in contact with the cladding, P_i is the pressure of the fission and fill gases inside the fuel rod.

$$P_i = \frac{R\Sigma M}{\Sigma V/T} \, , \qquad (2.141)$$

where

R = universal gas constant

$\Sigma V/T$ = sum of ratios of each free volume in pin to average absolute temperature

ΣM = total moles of released fission gas and initial filling gas.

Evaluation of the quantity of total fission gas release requires an estimation of gas released from each radial node at a number of axial elevations. Once convergence on surface temperature at a given axial elevation is obtained, the process must be repeated at a number of other axial elevations. Fission gas release must then be determined. If significant changes in internal pressure or gas thermal conductivity are found, the process must then be repeated at all elevations.

Cladding creep will not occur when the cladding is in a hydrostatic stress state; i.e., $\sigma_\theta = \sigma_r = \sigma_z$. According to the Prandtl-Reuss assumption, creep is determined by deviatoric stress, s. In the circumferential direction,

$$s_\theta = \sigma_\theta - \sigma_m \quad , \qquad (2.142)$$

where σ_m = mean stress = $(\frac{1}{3})(\sigma_r + \sigma_z + \sigma_\theta)$. To relate the multiaxial situation to the uniaxial condition in which strain rates are normally measured, an equivalent stress, σ_{eq}, is defined as

$$\sigma_{eq} = (1/\sqrt{2})[(\sigma_r - \sigma_\theta)^2 + (\sigma_\theta - \sigma_z)^2 + (\sigma_r - \sigma_z)^2]^{1/2} \quad . \qquad (2.143)$$

The equivalent stress is proportional to the total shear stress, which gives a measurement of total plastic creep deformation in a polycrystalline material. The value of σ_{eq} is substituted in an appropriate uniaxial creep equation, e.g., Eq. (2.75), to determine an equivalent creep rate $\dot{\epsilon}_c^{eq}$. Creep rates in the principal axes are then obtained from

$$\dot{\epsilon}_c^r/s_r = \dot{\epsilon}_c^\theta/s_\theta = \dot{\epsilon}_c^z/s_z = \left(\frac{3}{2}\right)(\dot{\epsilon}_c^{eq}/\sigma_{eq}) \quad . \qquad (2.144)$$

Hence, the tangential creep rate, $\dot{\epsilon}_c^\theta$, is obtained from

$$\dot{\epsilon}_c^\theta = \frac{3}{2} s(\dot{\epsilon}_c^{eq}/\sigma_{eq}) = (\epsilon^{eq}/\sigma_{eq})\left[\sigma_\theta - \frac{1}{2}(\sigma_r + \sigma_z)\right] \quad . \qquad (2.145)$$

Total creep strain during the time interval can then be estimated from

$$\epsilon_c = \dot{\epsilon}_c(\Delta t) \quad .$$

The procedure is satisfactory as long as small creep strains are obtained. If large creep strains are encountered, the time step must be reduced.

Note that radial stress, σ_r, is generally quite small in comparison with circumferential stress. If it is also concluded that interaction with the pellet stack largely eliminates axial cladding strain, then axial stress is relatively small. If both σ_r and σ_z are considered negligible, the equivalent stress, σ_{eq}, of Eq. (2.143) reduces to σ_θ. The circumferential creep rate from Eq. (2.145) then reduces to the creep rate from the appropriate uniaxial creep rate equation with σ_θ as the uniaxial stress. This simplified formulation is used in several design codes.

If the total cladding strain computed from Eq. (2.137), using inner pressure as gas pressure, leads to the conclusion that cladding diameter is equal to or less than the uncracked diameter of the pellet, then the fuel and cladding are in contact. The fuel will elastically strain the cladding so the pellet and cladding are just in contact. The elastic expansion, ΔD_i, of the cladding is

$$\Delta D_i = D_P - D_{i0}[\alpha_c(\Delta T) + \epsilon_c] \quad , \qquad (2.146)$$

where

D_{i0} = internal diameter of cladding at BOL conditions

D_P = uncracked diameter of hot pellet, including swelling effects.

Internal pressure P_i, required to obtain this expansion, can be computed from Eq. (2.37). As previously noted, contact pressure equals P_i less internal gas pressure. The contact pressure is then used to evaluate gap conductance. Again, an iterative procedure is required to obtain agreement between the assumed and calculated cladding and fuel movement.

When the central fuel temperature is in excess of $\sim 1750°C$ (some estimates place this temperature as high as 2000°C), the central fuel region contains columnar grains of 98 to 99% theoretical density. This is surrounded by a region where temperatures are ~ 1450 to 1750°C with equiaxial grains with densities of $\sim 97\%$ theoretical density. It was previously noted that reduction of voids in columnar and equiaxial grain regions led to a central hole. In addition, the fuel can no longer be considered incompressible; fuel creep and hot pressing should be considered.

Cheng and Ma[135] suggest a three-region approach. They determine the temperature at the edge of the columnar grain region by using Nichols' equation [Eq. (2.20)], in which they substitute the average temperature gradient (the gradient taken at one-third the fuel rod radius) for dT/dr. They assume the threshold temperature, T_ϕ, for the columnar region will allow motion of one-third of the fuel rod radius in the given irradiation time, t. To obtain the temperature at the edge of the equiaxial region, they assume that the rate of equiaxial grain growth is governed by the Lyons et al. equation[136]

$$D^3 - D_0^3 = k_0 t \exp(-Q/RT) \quad , \tag{2.147}$$

where

D_0 = initial grain size

D = grain size after annealing time, t

k_0, Q = constants

T = absolute temperature

t = irradiation time

R = universal gas constant.

Further, they assume that D, the grain diameter at the edge of the equiaxial region, is 15 μm. Thus, for any given irradiation time t, the temperature at the edge of the equiaxial region can be located.

With a known pellet surface temperature, outer radius, and value for T_2, the inner radius of the undisturbed fuel region is obtained by using Eq. (2.136). This radius and the values of T_1 and T_2 are then used in Eq. (2.136) to determine the outer radius of the columnar grain growth region. Once the radius of the columnar and equiaxial regions are known, the volume of the central hole can be estimated by assuming the reduction in void volume in these regions is equal to the volume of the central hole. The radius of the central hole then allows computation of the center fuel temperature by using Eq. (2.136) with the inner and outer radii of the columnar grain region. Note that an iterative process is again required since the surface temperature of the pellet is not known and must be determined from gap conductance. In turn, this depends on fuel thermal expansion, swelling, cladding expansion, etc.

A number of simplifying assumptions have been made about fuel behavior in the various zones. Perhaps the simplest is from Sayles[137] in which the outer unrestructured region is taken as nonplastic and brittle. Thermal expansion and swelling are assumed to cause this region to move outward. The inner regions are assumed plastic and restrained from outward motion. Thermal expansion in this region moves the material inward. In these inner regions, swelling is considered negligible.

Sun and Okrent[74] suggest a somewhat more sophisticated approach. They assume that in the columnar grain growth region, plastic flow is large and this region will simply act to transmit gas pressure in the central hole to the remaining fuel. However, the remaining fuel is considered to deform elastically and inelastically due to creep, swelling, and hot pressing. By using analytical expressions for each of the deformation mechanisms and substituting these expressions into equations describing the strain-stress relations of the fuel cylinder, deformations are determined.

2-4.2 Computer Programs for Fuel Element Design

The complexity of the calculations required to treat fuel and cladding deformation and restructuring, as well as fission gas release, obviously makes it desirable to perform these calculations via computer. One of the more sophisticated approaches to this problem has been the CYGRO code.[134,138] Both fuel and cladding are divided into a series of concentric rings linked by continuity laws of force and displacement. The stresses, strains, and all properties are taken to be uniform throughout a given ring. During each time increment, strain rates and stress are taken as constraints in a given ring. Strain rates and stress are related by Prandtl-Reuss equations that are modified to include the effects of fuel growth, thermal expansion, plasticity, and creep. The size of a time step is limited so changes in stress and strain are small.

Fuel swelling is computed via Nichols and Warner's BUBL model.[89] Porosity migration is treated by Nichols' method.[27] The total solid volume in each ring is held constant so that the radii of the boundaries follow the fuel. The heat generation rate in each ring is a constant fraction of the total heat generation throughout time history.

A comprehensive stress-strain analysis of the fuel is conducted. In the low-temperature region, allowance is made for the fact that if tensile stress exceeds a predetermined amount, the fuel cracks and tensile stress drops to near zero. Cracks are assumed to heal during compression. Frictional interaction between fuels and cladding is also considered. The finite element technique is used to solve stress-strain relations.

Although CYGRO conducts a very comprehensive analysis of the stress-strain relationship in the fuel, it considers only a single axial node. Fission gas pressure information as a function of time must come from another calculation.

The LIFE code[139] eliminates the need for obtaining fission gas release separately, since it is capable of considering a series of axial nodes at different power levels. This code was originally developed for fast reactor fuel, but versions satisfactory for use with LWR fuel are available.[140] The stress-strain analysis performed by the LIFE code is less elaborate than that done by CYGRO, since only three fuel regions (columnar, equiaxial, and unrestructured) are considered in the stress-strain analysis. The swelling and fission gas release analysis is more sophisticated since the GRASS code[90] is used for computation of these quantities and a number of axial nodes are considered.

Both CYGRO and LIFE are time consuming to run. A number of more rapidly running codes are available. In these codes, approximate stress analyses are conducted and swelling and fission gas release are described by analytical expressions. The COMETHE (Ref. 85), FRAP-S (Ref. 141), and GAPCON (Ref. 142) codes are examples of this approach. In all of these codes, a series of axial segments is considered so that a reasonable estimate of overall fission gas release can be obtained and its effect on cladding and fuel behavior can be determined.

The CYGRO code can consider a power distribution that is circumferentially nonuniform. Most fuel design codes utilize circumferentially symmetric models.

Olson and Boman[143] point out that in addition to circumferential variations, three-dimensional effects can arise: axial pellet separation, small pieces missing from the as-built pellets, azimuthal variations in the pellet-cladding clearance, and locally increased thermal impedance on the fuel rod surface. These effects were studied using the TAPA program.[144]

Olson and Boman[143] concluded that the maximum fuel rod temperature is increased in fuel pellets separated by an axial gap and in fuel pellets with a small missing segment. Stored energy is increased by steady operation by an axial pellet separation, but there is no effect on the peak cladding temperature during a hypothetical LOCA. None of the other effects resulted in a change in maximum stored energy or fuel temperature.

2-5 TRANSIENT FUEL ROD BEHAVIOR

2-5.1 Rapid Temperature Transients in Fuel Rods

2-5.1.1 Analytical Techniques

Under transient conditions when heat storage must be considered, the general conduction equation takes the form

$$c_p \rho \frac{\partial T}{\partial t} = \nabla(k \nabla T) + q''' \quad , \tag{2.148}$$

where t represents time, c_p specific heat, ρ density, and other symbols have their previous meanings.

We consider a fuel pellet of radius a surrounded by cladding with an outside radius b: If we ignore axial conduction and assume that k for the fuel is a constant, for the pellet region we write

$$k_1 \left(\frac{\partial^2 T_1}{\partial r^2} + \frac{1}{r} \frac{\partial T_1}{\partial r} \right) + q''' = c_{p_1} \rho_1 \frac{\partial T_1}{\partial t} \quad , \quad (0 < r \leqslant a) \quad , \tag{2.149}$$

while for the cladding region

$$k_2 \left(\frac{\partial^2 T_1}{\partial r^2} + \frac{1}{r} \frac{\partial T_2}{\partial r} \right) = c_{p_2} \rho_2 \frac{\partial T_2}{\partial r} \quad , \quad (a \leqslant r \leqslant b) \quad . \tag{2.150}$$

We consider a step change in heat transfer coefficient at the outer cladding at $t = 0$ and a subsequent reduction in heat generation rate. This corresponds to the situation arising during a transient when the fuel element is suddenly blanketed by steam [departure from nucleate boiling (DNB) occurs (Sec. 4-3.2)] and the film coefficient is drastically reduced. Heat generation rate is reduced on reactor scram.

A set of dimensionless groups is defined as follows:

$$B_1 = \frac{h_g b}{k_2} \quad , \quad B_2 = \frac{hb}{k_2} \quad , \tag{2.151}$$

where h_g is gap conductance and h is the film coefficient at rod surface

$$R_1 = \frac{r}{b} \quad , \quad R_p = \frac{a}{b} \quad , \quad \tau = \left(\frac{k_2}{c_p \rho_2}\right)\frac{t}{a^2} \quad , \tag{2.152}$$

$$C_p(R_1) = \frac{c_p(r)\rho(r)}{c_{p_2}\rho_2} \quad , \qquad K(R_1) = \frac{k_1(r)}{k_2} \quad , \tag{2.153}$$

$$Q(\tau) = \frac{q'''(t)}{q'''(0)} \quad , \qquad T^+ = \frac{k_2 T}{q'''(0)a^2} \quad . \tag{2.154}$$

Then Eqs. (2.149) and (2.150) can be written as

$$K\left(\frac{\partial^2 T_1^+}{\partial R_1^2} + \frac{1}{R_1}\frac{\partial T_1^+}{\partial R_1}\right) + Q = C_p \frac{\partial T_1^+}{\partial \tau} \quad , \quad (0 \leqslant R_1 \leqslant R_p) \quad , \tag{2.155}$$

$$\frac{\partial^2 T_2^+}{\partial R_1^2} + \frac{1}{R_1}\left(\frac{\partial T_2^+}{\partial R_1}\right) = \frac{\partial T_2^+}{\partial \tau} \quad , \qquad (R_p < R_1 \leqslant 1) \quad . \tag{2.156}$$

The boundary conditions are

$$B_1(T_1^+ - T_2^+) = -K \frac{\partial T_1^+}{\partial R_1} \quad , \quad (\text{at } R_1 = R_p) \quad , \tag{2.157}$$

$$B_1(T_1^+ - T_2^+) = \frac{\partial T_2^+}{\partial R_1} \quad , \quad (\text{at } R_1 = R_p) \quad , \tag{2.158}$$

$$B_2(T_2^+ - T_\infty^+) = -\frac{\partial T_2^+}{\partial R_1} \quad , \quad (\text{at } R_1 = 1) \quad , \tag{2.159}$$

where T_∞^+ is the dimensionless bulk coolant temperature.

These equations can be solved by a Hankel transform[145] or a Finite Integral transform.[146] In both cases, the exact solutions are in terms of infinite series. By using a finite integral transform technique, Matsch[146] obtained the following solution:

$$T^+(R_1,\tau) = \sum_n C_n(\tau)X_n(r) \quad , \tag{2.160}$$

where values X_n are obtained by considering an associated equation of the form

$$\nabla K(R_1)\nabla X(R_1) + (\lambda_n)^2 X(R_1) = 0 \quad , \tag{2.161}$$

to which the solutions are a set of eigenfunctions (X_n) true for discrete values of $(\lambda_n)^2$.

Thus, for the pellet we have

$$T_1^+(R_1,\tau) = \Sigma C_n'(\tau)X_n' \quad , \tag{2.162}$$

and for the cladding

$$T_2^+(R_1,\tau) = \Sigma C_n''(\tau)X_n'' \quad . \tag{2.163}$$

Matsch has shown that the eigenfunctions are given by

$$X_n' = J_0\left(\frac{\lambda_n R_1}{\sqrt{K'}}\right) \tag{2.164}$$

$$X_n'' = C_{1n} J_0(\lambda_n R_1) + C_{2n} Y_0(\lambda_n R_1) \ , \tag{2.165}$$

where J_0 is a Bessel function of the first kind, zero order, and Y_0 is a Bessel function of the second kind, zero order.

Eigenvalues λ_n are obtained from the determinantal equation:

$$
\begin{vmatrix}
\left[J_0\left(\dfrac{\lambda_n R_p}{\sqrt{K'}}\right) - \dfrac{\lambda_n\sqrt{K'}}{B_1} J_1\left(\dfrac{\lambda_n R_p}{\sqrt{K'}}\right)\right] & Y_0(\lambda_n R_p) \\[2ex]
\lambda_n\sqrt{K'} J_1\left(\dfrac{\lambda_n R_p}{\sqrt{K'}}\right) & Y_1(\lambda_n R_p)
\end{vmatrix}
\left[J_0(\lambda_n) - \dfrac{\lambda_n}{B_2} J_1(\lambda_n)\right]
$$

$$
+
\begin{vmatrix}
J_0(\lambda_n R_p) & \left[J_0\left(\dfrac{\lambda_n R_p}{\sqrt{K'}}\right) - \dfrac{\lambda_n\sqrt{K'}}{B_1} J_1\left(\dfrac{\lambda_n R_p}{\sqrt{K'}}\right)\right] \\[2ex]
J_1(\lambda_n R_p) & \lambda_n\sqrt{K'} J_1\left(\dfrac{\lambda_n R_p}{\sqrt{K'}}\right)
\end{vmatrix}
\left[Y_0(\lambda_n) - \dfrac{\lambda_n Y_1(\lambda_n)}{B_2}\right] = 0
\tag{2.166}
$$

where K' is the value of K in the pellet region (assumed constant) and J_1 is a Bessel function of the first kind, first order.

Once the values of λ_n are ascertained, constants C_{1n} and C_{2n} can be evaluated from

$$
C_{1n} = \frac{
\begin{vmatrix}
\left[J_0\left(\dfrac{\lambda_n R_p}{\sqrt{K'}}\right) - \dfrac{\lambda_n\sqrt{K'}}{B} J_1\left(\dfrac{\lambda_n R_p}{\sqrt{K'}}\right)\right] & Y_0(\lambda_n R_p) \\[2ex]
\lambda_n\sqrt{K'} J_1\left(\dfrac{\lambda_n R_p}{\sqrt{K'}}\right) & Y_1(\lambda_n R_p)
\end{vmatrix}
}{
\begin{vmatrix}
J_0(\lambda_n R_p) & Y_0(\lambda_n R_p) \\
J_1(\lambda_n R_p) & Y_1(\lambda_n R_p)
\end{vmatrix}
}
\tag{2.167}
$$

$$
C_{2n} = \frac{
\begin{vmatrix}
J_0(\lambda_n R_p) & \left[J_0\left(\dfrac{\lambda_n R_p}{\sqrt{K'}}\right) - \dfrac{\lambda_n\sqrt{K'}}{B_1} J_1\left(\dfrac{\lambda_n R_p}{\sqrt{K'}}\right)\right] \\[2ex]
J_1(\lambda_n R_p) & \lambda_n\sqrt{K'} J_1\left(\dfrac{\lambda_n R_p}{\sqrt{K'}}\right)
\end{vmatrix}
}{
\begin{vmatrix}
J_0(\lambda_n R_p) & Y_0(\lambda_n R_p) \\
J_1(\lambda_n R_p) & Y_1(\lambda_n R_p)
\end{vmatrix}
}
\ . \tag{2.168}
$$

Matsch obtains the values of C_n using the Laplace transformation for solution of N simultaneous differential equations. To do so, he defines five additional quantities

$$P_n = 2\pi \int_0^1 (X_n)^2 R_1 dR_1 \tag{2.169}$$

$$H_{n,m} = 2\pi \int_0^1 C_p(R) X_n X_m R_1 dR_1 \tag{2.170}$$

$$A_n = 2\pi \int_0^{R_1 P} X_n R_1 dR_1 \tag{2.171}$$

$$U_n = 2B_2(X_n)_{R=1} \tag{2.172}$$

$$\overline{V}_n = \sum_{n=1}^{n} Q_{m,n} \frac{\int_0^1 T^+_{(R,0)} X_n R_1 dR_1}{\int_0^1 (X_n)^2 R_1 dR_1} + T^+_\infty V_m + A_m \overline{Q}(\tau) \;, \tag{2.173}$$

where $\overline{Q}(\tau)$ is the Laplace transform of $Q(\tau)$.

The Laplace transform of $C_n(\tau)$ is then obtained by solving the determinantal equation

$$\overline{C}_n = \frac{\begin{vmatrix} [(H_{1,1})s + (\lambda_1)^2 P_1] & \cdots & \overline{V}_1 & \cdots & (H_{1,N})s \\ \cdots & \cdots & \cdots & \cdots & \cdots \\ (H_{1,n})s & \cdots & \overline{V}_n & \cdots & (H_{n,N})s \\ \cdots & \cdots & \cdots & \cdots & \cdots \\ (H_{1,N})s & \cdots & \overline{V}_N & \cdots & [(H_{N,N})s + (\lambda_N)^2 P_N] \end{vmatrix}}{\begin{vmatrix} [(H_{1,1})s + (\lambda_1)^2 P_1] & \cdots & (H_{1n})s & \cdots & (H_{1,N})s \\ \cdots & \cdots & \cdots & \cdots & \cdots \\ (H_{1,n})s & \cdots & [(H_{n,m})s + (\lambda_n)^2 P_n] & \cdots & (H_{n,N})s \\ \cdots & \cdots & \cdots & \cdots & \cdots \\ (H_{1,N})s & \cdots & (H_{n,N})s & \cdots & [(H_{n,m})s + (\lambda_N)^2 P_N] \end{vmatrix}} . \tag{2.174}$$

After solving these (N) determinants, the values of C_N are transformed from functions of (s) to functions of (τ) by using the inverse Laplace transform. Values of $C_n(\tau)$ and X_n are then substituted into Eqs. (2.162) and (2.163) to provide the desired result.

Matsch found that only the first three eigenvalues are needed to obtain the accuracy required by most calculations.

2-5.1.2 Finite Difference Procedure

In the case of very severe transients, the foregoing analytical techniques may not provide the desired accuracy due to large changes in fuel temperature and coolant conditions. This fact, coupled with the obvious complexity of the analytical procedures and the availability of digital computers, has led to an increased use of numerical techniques. Discrete time and space intervals are considered and the general conduction equation, Eq. (2.148), is rewritten in finite difference form. If axial conduction is ignored, for an annular region at a given elevation we write

$$V_j c_P(T_{j,i-1}) \times \left(\frac{T_{j,i} - T_{j,i-1}}{\Delta t_i} \right) = q_j' + \frac{T_{j-1,i-1} - T_{j,i-1}}{R_{j-1,j}} - \frac{T_{j,i-1} - T_{j+1,i-1}}{R_{j,j+1}} \quad, \qquad (2.175)$$

where

V_j = volume of region j, per unit of length

$c_P[T_{j,i-1}]$ = volumetric heat capacity of pellet at the temperature of region j at the end of time step $i - 1$

$T_{j,i}$ = temperature of region j at the end of time step $i - 1$

Δt_i = time increment

q_j' = heat generation per unit of time and length in region j

$R_{j-1,j}$ = resistance between region $j - 1$ and j computed as

$$\frac{1}{R_{j-1,j}} = \frac{k_p[T_{(j-1,j)i-1}](A_{j-1,j})}{r_j - r_{j-1}} \quad, \qquad (2.176)$$

with

$k_p[T_{(j-1,j),i-1}]$ = conductivity of UO_2 at a temperature that is the average temperature of regions $j - 1$ and j at the end of time step $i - 1$

$A_{j-1,j}$ = heat transfer area between regions $j - 1$ and j

r_{j-1} and r_j = average radii of these regions.

For the last region in contact with the cladding, we write

$$\frac{1}{R_{j,c}} = h_g \times A_{j,c} \quad, \qquad (2.177)$$

where h_g is contact conductance.

The basic equation can be solved as

$$T_{j,i} = \frac{q_j' \Delta t_i}{V_j c_P(T_{j,i-1})} + m_{j-1,j} T_{j-1,i-1} + m_{j,j+1} T_{j+1,i-1}$$

$$+ (1 - m_{j-1,j} - m_{j,j+1}) T_{j,i-1} \quad, \qquad (2.178)$$

where

$$m_{j-1,j} = \frac{\Delta t_i}{V_j c_P(T_{j,i-1}) R_{j-1,j}} \quad, \qquad (2.179)$$

$$m_{j,j+1} = \frac{\Delta t_i}{V_j c_P(T_{j,i-1}) R_{j,j+1}} \ . \tag{2.180}$$

The foregoing formulation is called an "explicit" solution since temperatures at new time i can be calculated directly (explicitly) given the conditions at time $i-1$. Equations of this nature are subject to numerical instability if the time step chosen is too large. Thus, Eq. (2.178) yields a negative (unstable) value for $T_{j,i}$ if

$$(m_{j-1,j} + m_{j,j+1} - 1)T_{j,i-1} \geqslant \frac{q'_j \Delta t_i}{V_j c_P(T_{j,i-1})} + m_{j-1,j}T_{j-1,i-1} + m_{j,j+1}T_{j+1,i-1} \ . \tag{2.181}$$

Since the right side of Eq. (2.181) is always positive, we can avoid this condition by assuring ourselves that

$$m_{j-1} + m_{j,j+1} - 1 \leqslant 0 \ . \tag{2.182}$$

Thus, the maximum value of Δt_i can be determined as

$$\Delta t_i \leqslant \frac{V c_P(T_{j,i-1})(R_{j-1,j})(R_{j,j+1})}{R_{j-1,j} + R_{j,j+1}} \ . \tag{2.183}$$

The minimum of Δt_i values determined for the several segments is used.

There are two disadvantages to this procedure: First, we are computing the net heat flow into a given annular segment based on temperatures at the previous time. Second, restriction on the size of the time increment may require an excessive amount of computations for problems extending over long time periods. These difficulties can be obviated using the "implicit" form of the difference equation. Here we determine heat flows from temperature differences at the advanced position in time. Thus, Eq. (2.175) becomes

$$V_j c_P \left(\frac{T_{j,i} + T_{j,i-1}}{2}\right) \times \frac{(T_{j,i} - T_{j,i-1})}{\Delta t_i} = q'_j + \frac{(T_{j-1,i} - T_{j,i})}{R_{j-1,j}} - \frac{(T_{j,i} - T_{j+1,i})}{R_{j,j+1}} \ . \tag{2.184}$$

Initially, the average specific heat can be assumed equal to its value at the previous time. We then have a set of linear equations that can be solved simultaneously for $T_{j,i}$. If specific heat changes significantly with temperature, an iteration can be required using updated values of c_p.

While the implicit method removes tight restrictions on the value of Δt_i chosen, use of a large value of Δt_i increases the error due to discretization. This discretization error is of $O[\Delta t_i + (\Delta r)^2]$.

The foregoing finite difference procedures can also be used for calculating cladding temperatures by using cladding properties and setting q'_j to zero at interior nodes and equal to the heat generated by the zirconium-water reaction in the outermost region. For the node point adjacent to fuel, R_j is defined in terms of gap conductance, while for the node point adjacent to the coolant, R_j is defined in terms of coolant heat transfer coefficient.

Finite difference procedures for transient fuel element temperature calculations are now generally incorporated into the computer codes used for analyzing core thermal and hydraulic transients (e.g., the COBRA-IIIC code[147] or later models). Power generation versus time is input, and the program computes cladding temperature and heat input to the coolant.

2-5.2 Simplified Analytical Techniques for Temperature Transients

Analytical techniques, which are less tedious than described in Sec. 2-5.1, can be useful in some instances. They can provide a rapid means for obtaining approximate estimates under many conditions. In addition, for relatively slow transients, the results obtained can be adequate and not require a more sophisticated procedure. The loss-of-flow accident (LOFA) with scram sometimes falls into this category. For a UO_2 fuel rod, maximum cladding temperature occurs several seconds after initiation of the accident and a lumped parameter procedure yields a reasonable estimate.

2-5.2.1 The Lumped Parameter Technique

Lumped parameter techniques are among the earliest procedures used for examining temperature transients.[148] In the lumped parameter procedure suggested by Tong,[149] thermal resistances and the capacitances of the pellet and cladding are evaluated at their average condition in time and space. Each quantity is lumped at the middle of the physical geometry and axial conduction is neglected. Again, we consider the situation where there is a step reduction in the coolant heat transfer coefficient and a subsequent reduction in heat generation rate due to reactor scram.

Rate equations for heat transfer from the UO_2 fuel rod can be written as

$$q'_n = C_1 \frac{dT_1}{dt} + \frac{T_1 - T_2}{R_1} \ , \tag{2.185}$$

and

$$\frac{T_1 - T_2}{R_1} = C_2 \frac{dT_2}{dt} + \frac{T_2 - T_c}{R_2} \ , \tag{2.186}$$

where

q'_n = nuclear heating, Btu/s ft of fuel rod

C_1 = thermal capacitance of pellet, Btu/°F ft, $C_1 = \pi r_1^2 c_{p1} \rho_1$

C_2 = thermal capacitance of clad, Btu/°F ft, $C_2 = 2\pi r_2 (\Delta r) c_{p2} \rho_2$ for thin cladding

R_1 = resistance of UO_2 and gap, s ft °F/Btu

= $[1/(8\pi k_1)] + [1/(2\pi r_1 h_g)]$, where k_1 is UO_2 thermal conductivity and h_g is gap conductance

R_2 = resistance of coolant film = $1/(2\pi r_2 h)$, s ft °F/Btu

T_1 = average pellet temperature, °F

T_2 = average cladding temperature, °F

T_c = bulk coolant temperature, °F.

After a pipe rupture, the system pressure and saturation temperature of the coolant drop with time. Hence, $T_c = T_c(t)$. Time t is counted from the instant of rupture. From the Laplace transformation of Eqs. (2.185) and (2.186), we get

$$T_2(s) = \frac{R_2 q_n'(s) + (C_1 R_1 s + 1)T_c(s) + R_2 C_1 T_1(0) + R_2 C_2 (R_1 C_1 s + 1)T_2(0)}{R_1 R_2 C_1 C_2 s^2 + (R_1 C_1 + R_2 C_2 + R_2 C_1)s + 1} , \quad (2.187)$$

$$T_1(s) = \frac{\left(R_2 C_2 s + 1 + \dfrac{R_2}{R_1}\right) R_1 q_n'(s) + T_c(s) + \left(R_2 C_2 s + 1 + \dfrac{R_2}{R_1}\right) R_1 C_1 T_1(0) + R_2 C_2 T_2(0)}{R_1 R_2 C_1 C_2 s^2 + (R_1 C_1 + R_2 C_2 + R_2 C_1)s + 1} .$$

$$(2.188)$$

By knowing coolant temperature and fuel rod power as a function of time, the histories of pellet and cladding temperatures can be computed from the inverse transformation of Eqs. (2.187) and (2.188).

In the analysis of loss-of-flow transient, two simplifications can be made: First, when the pumps lose power, system pressure does not change significantly and coolant temperature T_c remains approximately constant; second, maximum cladding temperature usually occurs within 10 s of the instant of loss of power to the pump. Hence, decay heat can be assumed constant during this short time period. Our boundary conditions now become

for $t \leqslant 0$: $q_n' = q_n'(0)$, $R_2 = R_{2,0}$

for $0 < t \leqslant t_1$: $q_n' = q_n'(0)$, $R_2 = R_{2, \text{film boiling}}$

for $t \geqslant t_1$: $q_n' = \beta q_n'(0)$, $R_2 = R_{2, \text{film boiling}}$.   (2.189)

By taking advantage of the constancy of T_c, we can solve Eqs. (2.186) and (2.187) for T_i and then differentiate with respect to t to obtain dT_1/dt. We then rewrite Eq. (2.185) as

$$q_n' = C_1 C_2 R_1 \frac{d^2\theta}{dt^2} + \left(C_1 + C_2 + \frac{C_1 R_1}{R_2}\right)\frac{d\theta}{dt} + \frac{\theta}{R_2} , \quad (2.190)$$

where $\theta = T_2 - T_c$.

When the Laplace transform is taken, we obtain

$$\frac{1}{s}[1 - \beta \exp(t_1 s)] = C_1 C_2 R_1 \left[\bar{\theta} s^2 - \theta(0)s - \frac{d\theta(0)}{dt} \right] + \left(C_1 + C_2 + \frac{C_1 C_2}{R_2} \right)$$

$$\times [\bar{\theta} s - \theta(0)] + \frac{\bar{\theta}}{R_2} \quad , \tag{2.191}$$

where $\bar{\theta}$ represents the Laplace transform of θ.

The value $\theta(0)$ is given by

$$\theta(0) = q'_n(0)R_{2,0} \quad , \tag{2.192}$$

and $d\theta(0)/dt$ from the relationship

$$T_1(0) = q'_n(0)(R_1 + R_{2,0}) \quad , \tag{2.193}$$

and Eq. (2.186). Substitution of these numerical values, followed by the inverse transformation, yields θ as a function of time.

When the heat transfer coefficient undergoes a step reduction at $t = 0$, and heat generation is reduced to 15% of the steady-state value at $t = t_1$, then Matsch obtains

$$\left. \begin{aligned} \theta &= 1.19 - 0.683 \exp(-0.466t) - 0.488 \exp(-4.79t) && (t < t_1) \\ \theta &= 0.177 - \exp(-0.466t)[0.683 - 1.121 \exp(0.466t_1)] \\ &\quad - \exp(-4.794t_1)[0.488 + 0.1108 \exp(4.79t_1)] && (t > t_1) \end{aligned} \right\} . $$

$$\tag{2.194}$$

Figure 2-22 shows comparison of this solution with the integral transform solution of the same problem.

Chen et al.[150] extended the lumped parameter technique to consider phase changes (cladding, melting, and fuel melting). They considered the problem of loss of flow at constant power (failure to scram) and developed equations for estimating the times at which cladding and fuel melting begin. Steady-state conditions are computed first, assuming the coolant is at saturation. The average cladding temperature, T_{c_O}, at steady state is

$$T_{c_O} = T_{\text{sat}} + \frac{q''' R_F^2}{2} \left[\frac{1}{h_0 R_c} + \frac{1}{k_c} \left(\frac{1}{2} - \frac{R_F^2}{R_c^2 - R_F^2} \ln \frac{R_c}{R_F} \right) \right], \tag{2.195}$$

and the average steady-state fuel temperature, T_{F_O}, is given by

$$T_{F_O} = T_{\text{sat}} + \frac{q''' R_F^2}{2} \left(\frac{1}{4k_F} + \frac{1}{h_0 R_c} + \frac{1}{h_{\text{gap}} R_F} + \frac{1}{k_c} \ln \frac{R_c}{R_F} \right), \tag{2.196}$$

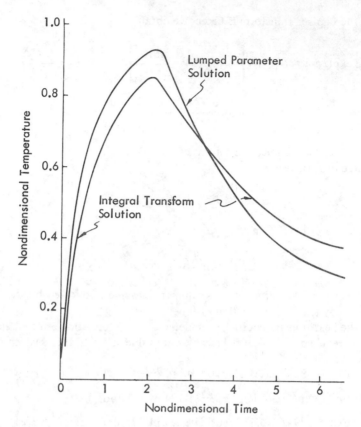

Fig. 2-22.   Comparison of solutions for fuel element transient by lumped-parameter and finite-integral transform technique. [From *Nucl. Sci. Eng.*, **11**, 340 (1961).]

where

h_0 = steady-state (pre-DNB) coolant heat transfer coefficient

q''' = volumetric heat generation rate

R_F = outer radius of fuel region

R_c = outer radius of cladding

k_c, k_F = thermal conductivity of cladding and fuel, respectively

T_{sat} = saturation temperature.

They then assume that the coolant heat transfer coefficient goes to zero (vapor blanketing) and remains constant at that value. After defining the following symbols, they develop the working equations shown in Table 2-VI, where

t_d = time at which vapor blanketing (DNB) occurs

ρ_c, C_{pc}, L_c = density, specific heat, and latent heat of cladding, respectively

ρ_F, C_{PF}, L_F = density, specific heat, and latent heat of fuel, respectively

$$h' = \left(\frac{R_F}{4k_F} + \frac{1}{h_{gap}} + \frac{R_c - R_F}{2k_c}\right)^{-1}, \quad \text{average overall heat transfer coeffi-}$$
cient

$$C_F = \frac{R_F \rho_F C_{PF}}{2h'}, \quad \text{calculated at the steady-state maximum fuel tempera-}$$
ture

$$C_F' = \frac{R_F \rho_F C_{PF}}{2h'}, \quad \text{calculated at average temperature } T_{av} = \frac{T_{FP} + T_F(t_m)}{2}$$

$$C_c = \frac{(R_c^2 - R_F^2)\rho_C C_{pc}}{2R_F h'}, \quad \text{calculated at the cladding melting temperature,}$$
T_{cp}

$$Q = \frac{R_F q'''}{2h'}.$$

2-5.2.2 Electrical Analog

The lumped parameter technique is basically equivalent to using an electrical analog. Conditions at the nodal points of the electrical network represent the average conditions of a particular region. To use such an analog, we note the correspondence between voltage and temperature, current and flow, and the product of electrical resistance and the capacity of the electrical system and thermal diffusivity. Thus, Fourier's law of $q = \Delta T/R_t$ is represented by Ohm's law for current flow $i = \Delta E/R$. The change in system thermal energy is given by

$$\frac{dQ}{dt} = c_p \rho V \frac{dT}{dt}, \tag{2.197}$$

where

Q = heat content, Btu

c_p = specific heat

V = volume

ρ = density.

The change in electrical energy is represented by

$$\frac{dQ_e}{dt} = C \frac{dE}{dt}, \tag{2.198}$$

where Q_e is electric charge on the condenser and C is condenser capacitance.

TABLE 2-VI

WORKING EQUATIONS FOR THE CHEN et al.[150] TRANSIENT CONDUCTION MODEL WITH PHASE CHANGE

Phases	Timing	Average Fuel and Cladding Temperatures
Transient phase before clad melting	$t_d < t \leqslant t_p$ (starting time of clad melting), where $$t_p = \begin{cases} t_d + \left(\dfrac{T_{cp} - T_{c0}}{T_{F0} - T_{c0}}\right) C_c & \text{for } t_p - t_d \leqslant \dfrac{C_F C_c}{C_F + C_c} \\[3ex] t_d + \left[\dfrac{T_{cp} - T_{c0}}{T_{F0} - T_{c0}} - \left(\dfrac{C_F}{C_F + C_c}\right)^2\right](C_F + C_c) & \text{for } t_p - t_d > \dfrac{C_F C_c}{C_F + C_c} \end{cases}$$	Case 1: for $t - t_d \leqslant \dfrac{C_F C_c}{C_F + C_c}$ $$T_c = T_{c0} + \frac{Q}{C_c}(t - t_d) \quad \text{and} \quad T_F = T_{F0}$$ Case 2: for $t - t_d > \dfrac{C_F C_c}{C_F + C_c}$ $$T_c = T_{c0} + \frac{Q}{C_F + C_c}\left(\frac{C_F^2}{C_F + C_c} + t - t_d\right)$$ $$T_F = T_{F0} - \frac{C_F C_c}{(C_F + C_c)^2}(T_{F0} - T_{c0}) + \frac{Q}{C_F + C_c}(t - t_d)$$
Clad melting phase	$t_p < t \leqslant t_m$ (complete time of clad melting) Case 1: for $t_p - t_d \leqslant \dfrac{C_F C_c}{C_F + C_c}$ $$t_m = \begin{cases} t_p + \dfrac{(R_c^2 - R_F^2)\rho_c L_c}{2 R_F h'(T_{F0} - T_{cp})} & \text{for } t_m - t_p \leqslant C_F \\[3ex] t_p + \dfrac{(R_c^2 - R_F^2)\rho_c L_c}{2 R_F h'(T_{F0} - T_{c0})} + C_F \dfrac{T_{cp} - T_{c0}}{T_{F0} - T_{c0}} & \text{for } t_m - t_p > C_F \end{cases}$$ Case 2: for $t_p - t_d > \dfrac{C_F C_c}{C_F + C_c}$	$T_c = T_{cp}$ (clad melting temperature) Case 1: for $t_p - t_d \leqslant \dfrac{C_F C_c}{C_F + C_c}$ $$T_F = \begin{cases} T_{F0} + (T_{cp} - T_{c0})\dfrac{t - t_p}{C_F} & \text{for } t - t_p \leqslant C_F \\[3ex] T_{F0} + T_{cp} - T_{c0} & \text{for } t - t_p > C_F \end{cases}$$ Case 2: for $t_p - t_d > \dfrac{C_F C_c}{C_F + C_c}$

Transient phase before fuel melting		

$$t_m = \begin{cases} t_p + \dfrac{(R_c^2 - R_F^2)\rho_c L_c}{2R_F h'(T_{F_0} - T_{c0})}\left(\dfrac{C_F + C_c}{C_c}\right), & \text{for } t_m - t_p \leq C_F \\[3ex] t_p + \dfrac{(R_c^2 - R_F^2)\rho_c L_c}{2R_F h'(T_{F_0} - T_{c0})} + \dfrac{C_F^2}{C_F + C_c}, & \text{for } t_m - t_p > C_F \end{cases}$$

$t_m < t \leq t_{PF}$ (starting time of fuel melting), where

$$t_{PF} = \begin{cases} t_m + \dfrac{[T_{FP} - T_F(t_m)]C_F'}{T_{F_0} - T_{c0} - T_F(t_m) + T_{cp}} & \text{for } t_{PF} - t_m \leq \dfrac{C_F' C_c}{C_F' + C_c} \\[3ex] t_m + \dfrac{C_c}{C_F' + C_c}\left[T_F(t_m) - T_{cp} - \dfrac{\left[T_F(t_m) - T_{c0} \right]}{(T_{F_0} - T_{c0})/(C_F' + C_c)} \right] & \text{for } t_{PF} - t_m > \dfrac{C_F' C_c}{C_F' + C_c} \end{cases}$$

$$T_F = \begin{cases} T_{cp} + \dfrac{T_{F_0} - T_{c0}}{C_F + C_c}(C_c + t - t_p), & \text{for } t - t_p \leq C_F \\[3ex] T_{F_0} + T_{cp} - T_{c0}, & \text{for } t - t_p > C_F \end{cases}$$

$$T_c = \begin{cases} T_{cp} + \dfrac{1}{C_c}[T_F(t_m) - T_{cp}](t - t_m), & \text{for } t - t_m \leq \dfrac{C_F' C_c}{C_F' + C_c} \\[3ex] T_{cp} + \dfrac{T_{F_0} - T_{c0}}{C_F' + C_c}\left(\dfrac{C_F'[T_F(t_m) - T_{cp}]}{T_{F_0} - T_{c0}} + t + t_m - \dfrac{C_F' C_c}{C_F' + C_c}\right) & \text{for } t - t_m > \dfrac{C_F' C_c}{C_F' + C_c} \end{cases}$$

$$T_F = \begin{cases} T_F(t_m) - \dfrac{1}{C_F'}[T_F(t_m) - T_{cp} - T_{F_0} + T_{c0}](t - t_m) & \text{for } t - t_m \leq \dfrac{C_F' C_c}{C_F' + C_c} \\[3ex] T_F(t_m) - \dfrac{T_{F_0} - T_{c0}}{C_F' + C_c}\left(\dfrac{C_c[T_F(t_m) - T_{cp}]}{T_{F_0} - T_{c0}} - t + t_m - \dfrac{C_c^2}{C_F' + C_c}\right) & \text{for } t - t_m > \dfrac{C_F' C_c}{C_F' + C_c} \end{cases}$$

Fuel melting phase		

$t_{PF} < t \leq t_{Fm}$ (complete time of fuel melting)

$$t_{Fm} = \begin{cases} t_{PF} + \dfrac{R_F \rho_F L_F}{2h'[T_{F_0} - T_{c0} - T_{FP} + T_c(t_{PF})]} & \text{for } t_{Fm} \leq t_{PF} + C_c \\[3ex] t_{PF} + \dfrac{R_F \rho_F L_F}{2h'(T_{F_0} - T_{c0})} + \dfrac{C_c[T_{FP} - T_c(t_{PF})]}{T_{F_0} - T_{c0}} & \text{for } t_{Fm} > t_{PF} + C_c \end{cases}$$

$$T_c = T_{FP} - [T_{FP} - T_c(t_{PF})]\exp\left(-\dfrac{t - t_{PF}}{C_c}\right)$$

$$T_F = T_{FP} \quad \text{(fuel melting temperature)}$$

As a simple illustration, consider the behavior of a fuel pellet in which (a) heat generation has ceased, (b) its outer radius is being cooled by a coolant at some constant temperature T_c, and (c) with a very high heat transfer coefficient. Since we are considering only the average temperature of the pellet, we write

$$C_1 \frac{dT}{dt} = -\frac{(T - T_c)}{R_1} \; , \tag{2.199}$$

where

C_1 = thermal capacitance of the pellet, Btu/°F ft, $C_1 = r_1^2 c_p \rho_1$

R_1 = thermal resistance of pellet, s ft °F/Btu.

The solution of this equation can be put in the form

$$q' = \frac{T_o - T_c}{R_1} \exp[-t/(R_1 C_1)] \; . \tag{2.200}$$

The equivalent electrical circuit is shown in Fig. 2-21a; we analogously obtain to Eq. (2.200) by

$$i = \frac{E_0}{R_i} \exp[-t/(R_1 C_1)] \; . \tag{2.201}$$

In equations of this form, which describe circuit transients, the denominator of the exponential (in this case $R_1 C_1$) is called the time constant of the element and its dimensions are in seconds.

We now consider the previous problem where heat generation rate is a function of time and both fuel rod and coolant thermal resistance must be considered. The equivalent electrical circuit is shown in Fig. 2-23. Note the direct correspondence between the circuit elements and thermal capacitances and resistances of Eqs. (2.185) and (2.186). Greater accuracy can be obtained by using additional nodal points.

The electrical analog technique is widely used for control analyses and is sometimes a useful tool for accident analyses since it allows a ready combination of both reactivity and thermal effects.[151] However, the most important application of electrical analogs is their use as plant simulators for reactor operator training.

2-5.3 Cladding Behavior During Transients

2-5.3.1 Behavior During Hypothetical LOCA

While many early transient analyses were concerned with demonstrating that center fuel melting did not occur, more recent studies are primarily concerned with cladding behavior. It is recognized that analytical techniques are valid only as long as a coolable geometry is maintained. During a LOCA, cladding temperature can rise to sufficiently high levels so that the hottest rods swell and burst locally with a longitudinal split. Providing the cladding is not too brittle, the cladding remains in

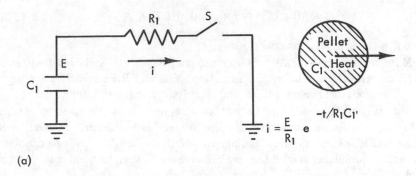

(a)

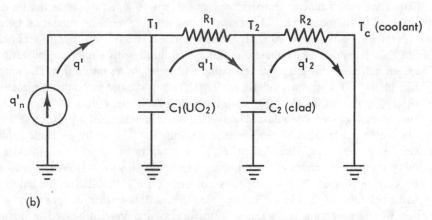

(b)

Fig. 2-23. Equivalent electrical circuits for fuel rod transients.

one piece and the pellets are restrained. The design limitations established are based on this requirement.

It was previously observed that Zircaloy reacts with steam at high temperatures to form zirconium oxide and hydrogen. The rate at which this reaction occurs is generally described by the well-known Baker-Just relation[152]

$$W^2 = (33.3 \times 10^6 \, t) \exp(-45\,000/RT) \,, \qquad (2.202)$$

where

W = weight of zirconium oxidized, mg/cm^2

T = temperature of unoxidized zirconium, K

t = time, s

R = universal gas constant, 1.987 cal/mole K.

The reaction rate can be related to the distance δ that the metal-metal oxide interface has moved from its original position by

$$d\delta/dt = (K/\delta)\exp(-45\,000/RT) , \qquad (2.203)$$

where K is the parabolic rate law constant, $0.3937\ cm^2\ s$.

More recent experiments by Grossman and Rooney[153] yield reaction rates in general agreement with Baker and Just. Grossman and Rooney's rates are slightly lower at temperatures below $1100°C$ and slightly higher above $1300°C$.

The zirconium oxide film formed on the surface of the cladding acts as a thermal barrier. At the same time, zirconium metal is consumed, effectively reducing the wall thickness of the cladding. In addition, Zircaloy tends to dissolve its own oxide, producing a cladding with increased oxygen content and decreased ductility.

Earlier we noted that at temperatures above $\sim 1040\ K$, alpha-phase zirconium is transferred to the beta phase. At temperatures where zirconium oxidation takes place (between 1400 and 1700 K) the outer surface is oxidized to ZrO_2. This is adjacent to a region of alpha-phase zirconium stabilized with a high concentration of oxygen and then a region of mixed alpha and beta phases inside this. The inner region is likely to be pure beta-phase material. The oxide and alpha-phase material are brittle; ductility of the cladding is primarily due to the beta phase.

The embrittlement of oxidized Zircaloy has been studied by Hobson and Rittenhouse[154] and Hobson.[155] Hobson and Rittenhouse[154] initially concluded that the ductility of Zircaloy depended on F_w, the fraction of original wall thickness remaining in the beta phase at operating temperature. However, Hobson[155] subsequently concluded that ductility depends not only on F_w, but also on the amount and distribution of oxygen in the beta phase. These latter quantities appear to be a function of the oxidizing temperature with increased oxygen dissolved at higher oxidizing temperatures. In view of these facts, the acceptance criteria for ECCS for water-cooled reactors[156] require that total oxidation of the cladding be limited to a maximum of 17% of the original cladding material and that the maximum cladding temperature during the transient be limited to $2200°F$. Therefore, analyses of a hypothetical LOCA must demonstrate that neither of these limits are exceeded.

Metallurgical interaction between Zircaloy and stainless-steel or nickel-based alloys used as spring or spacer grid materials can occur at temperatures in excess of $900°C$. The eutectic temperatures for the possible alloy systems[157] are: Zr-Fe, $934°C$; Zr-Ni, $961°C$; and Zr-Cr, $1300°C$. Alloying is apparently inhibited to some degree by crud and oxide layers. Pickman[157] notes that United Kingdom Atomic Energy Authority experiments show that Zircaloy-2 can braze to stainless-steel grid ferrules at temperatures in excess of $1200°C$. The $2200°F$ ($1204°C$) temperature limit set by U.S. emergency core cooling acceptance criteria effectively serves to limit possible alloying.

Cladding failure can occur when internal gas pressure causes cladding ballooning and subsequent rupture at local spots. It is of interest to be able to calculate the degree to which such failures are possible under various hypothetical conditions. Several computer code models, such as SST (Ref. 140) and FRAP-T (Ref. 158) have been designed to perform such calculations. Initial conditions are determined by a steady-state model that simulates the power-time history of the fuel rod to the point at which the transient begins. The computer codes then attempt to estimate fuel and cladding behavior.

In the central region of the fuel rod, the substantial zirconium-steam reaction will lead to rapid expansion of the cladding. The FRAP code considers that gas flows from the gas plenum to the expanded region through the fuel-cladding gap. Laminar flow is assumed in the gap. Gas pressures are computed as a function of axial location and time. External and internal pressures are used to determine stresses in the cladding. As cladding temperature rises, stresses can reach a level sufficient to cause ballooning at some location. The shape of the localized cladding bulge is taken as that of a rotated sine curve.

The equilibrium equation for a thin membrane element in the bulge area can be written as

$$(\sigma_a/r_a) + (\sigma_\theta/r_c) = P/L_c \ , \tag{2.204}$$

where

P = differential pressure

σ_a, σ_θ = axial and circumferential stresses, respectively

r_a, r_c = axial and circumferential radii of curvature, respectively

L_c = cladding thickness.

Significant deformation is not obtained until axial and radial stresses in the cladding exceed the yield stress or

$$P/L_c > \sigma_y \left(\frac{1}{r_a} + \frac{1}{r_c} \right) . \tag{2.205}$$

When this holds, the element is unstable and will deform. The cladding is then deformed at the unstable nodes to restore stability. At high strain rates (strain rates in excess of $10^{-7} \ s^{-1}$) and temperatures between 750 to 1050°C, cladding is assumed superplastic. Under such conditions, an increase in strain causes a large increase in stress. If $\dot{\epsilon}_{i-1}$ and σ_{i-1} are, respectively, strain rate and stress in the previous period, then stress σ_i in the current period and current strain rate, $\dot{\epsilon}_i$, are related by

$$\sigma_i = \sigma_{i-1}(\dot{\epsilon}_i/\dot{\epsilon}_{i-1})^m \ , \tag{2.206}$$

where m is a constant for the material.

The FRAP-T model is also capable of considering the situation where thermal expansion forces one or more fuel pellets outward against the cladding. The FRAP model also considers that some pellets below the level at which cladding-pellet interaction occurs can be prevented from expanding freely in the axial direction. The stresses so induced can be computed.

There has been some dispute about the criterion for cladding failure (rupture). Hannerz and Vesterlund[159] report data that indicate burst stress is a function of heating rate. However, the data of Hobson and Rittenhouse[160] seem to show very small effects of heating rate. The FRAP code assumes the normal strain failure theory applies (failure occurs when strain exceeds a preset value that is a function of temperature). Figure 2-24 illustrates the cladding failure criteria utilized by FRAP-T.

Significant swelling in the region near a spring can exceed the working range of the grid. This can jam the fuel rod against the grid preventing axial motion and inducing fuel rod bowing. The effects of ballooning can, however, be more significant. Interaction between adjacent ballooned rods can effectively block a channel. While such blockage does not have a deleterious effect at high flooding rates (rates in excess of one inch per minute), at lower rates the effect can be deleterious.[161] Current (1976) regulations conservatively require that at these low flooding rates (a) cooling calculations are based on the assumption that cladding temperatures are determined by steam cooling alone, and (b) that the effects of flow blockage are

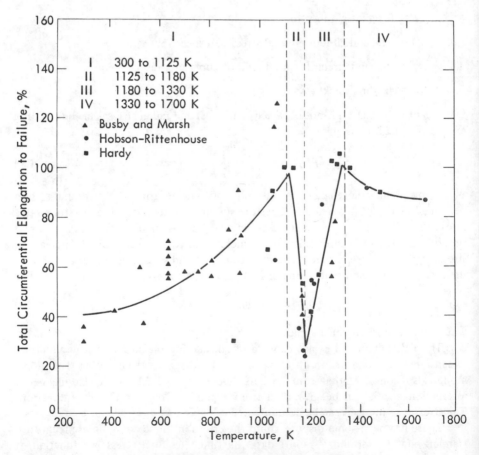

Fig. 2-24.   Cladding failure correlation (Ref. 50).

considered. Hall and Duffy[162] suggest a procedure for calculating the effect of cladding ballooning on coolant flow. They estimate the flow in the blocked region by assuming that flow must redistribute in such a manner that pressure drop across the blocked region must be equal to that of the unblocked region in a parallel channel.

2-5.3.2 Behavior During Power Shocks

It was previously observed that rapid increases in fuel rod power can cause cladding cracking and failure. When there is a zero gap between fuel and cladding, thermal expansion of UO_2 during a power increase applies a tensile stress to the cladding. The magnitude of the stress depends on the change in power level. At sufficiently high tensile stresses, chemically-assisted corrosion can lead to rapid cladding cracking and failure. Computer codes such as FRAP-S (Ref. 141) and GAPCON (Ref. 142) are capable of analyzing the reactor cycle and predicting cladding stress levels using nominal values of dimensions, properties, and conditions. Rolstad[163] argues that such computer programs are unsuitable for predicting power shock failures since rod failures in power reactors have been observed at power shocks less than the threshold determined by experiments in test reactors.

Rolstad[163] concludes that stochastic elements play an important role in determining power shock failures. He considers that failures occur when controlling parameters are at the most unfavorable ends of their tolerance bands, or when the most unfavorable fuel crack pattern coincides with the worst cladding defect. Therefore, he proposes a probabilistic approach. In Rolstad's POSHO code,[142] local power shocks resulting from planned fuel operation are calculated. These are then processed through a fuel failure probability routine. Failure probability estimates are based on fuel design parameters, burnup, and previous power shock failure experience. The failure probabilities for a given power change are a function of local power level. The number of cladding cracks produced by the change are obtained by multiplying the number of pellet interfaces at a given power level by the average failure probability at that power level.

Fuel failures due to power ramps were significant problems in the CANDU-type reactor where at-power refueling can impose rapid power changes on the fuel elements being moved. Studies of CANDU fuel elements have concluded that at a given burnup level there is a range of cladding stresses above which fuel failures are likely. The stress range at which failures begin decreases as burnup increases. It has been found that application of a colloidal dispersion of graphite to the inner wall of the Zircaloy cladding decreases the cladding's susceptibility to power ramp failures.[164] Since graphite application allows operation at increased stress levels without failures, use of this process has very significantly decreased fuel rod failures in CANDU reactors.

2-6 SUMMARY

Although a number of alloy fuels with reasonable in-reactor behavior have been developed, the superior performance of ceramic fuels allowed them to become the predominant fuel for water-cooled reactors. Furthermore, performance and cost advantages of Zircaloy have led to the use of Zircaloy as cladding in all recent PWR designs. Therefore, fuel element analysis is almost exclusively concerned with the behavior of pellets consisting largely of UO_2 clad in Zircaloy.

Early design approaches were generally based on idealized conditions that assumed a solid pellet with an annular gas gap between pellet and cladding. This picture is now recognized as overly simplistic. Thermal analyses now account for (a) fuel expansion, swelling, cracking, and restructuring, (b) cladding expansion, elastic and plastic strain, and creep, and (c) fission gas release. Computer programs capable of modeling this behavior and following the fuel element through a reactor cycle are available and in general use. Since fission gas pressure depends on behavior throughout the rod, programs capable of examining the entire fuel rod should be used.

The necessity to demonstrate that fuel behavior lies within the accepted criteria during hypothetical accident conditions has led to an extension of steady-state computer programs to encompass transient behavior. The stringency of safeguards criteria requires that hypothetical accident conditions be carefully considered.

REFERENCES

1. W. R. McDonnell and E. F. Sturcken, "Development of High Performance Uranium-Metal Fuels for Savannah River Reactor," *Nucl. Technol.*, **26**, 420 (1975).

2. C. R. Tipton, Ed., *Reactor Handbook, Vol. I, Materials*, 2nd ed., p. 193, Interscience Publishers, Inc., New York (1960).

3. C. R. Tipton, Ed., *Reactor Handbook, Vol. I, Materials*, 2nd ed., pp. 195-196, Interscience Publishers, Inc., New York (1960).

4. J. A. Taub, "Fabrication of Uranium-Molybdenum Alloy Fuels for Pulsed Reactors," *Nucl. Technol.*, **28**, 77 (1976).

5. R. W. Dayton, "Objectives of U. S. Research in Nuclear Fuels for Civilian Power Reactors," *Proc. Third Int. Am. Symp. Peaceful Application of Nuclear Energy*, Pan American Union, Washington, D. C. (1961).

6. C. Collet, J. P. Lanque, and M. Columbie, "Study of Mechanical and Thermal Properties of Uranium Alloys with Small Additions," *Mem. Sci. Rev. Metall.*, **64**, 393 (Apr. 1967).

7. F. W. Albough, S. H. Bush, J. J. Cadwell, D. R. deHales, and D. C. Woster, Battelle Northwest Laboratory Communication appearing in *Reactor Materials*, **10**, *3*, 136 (1967).

8. M. A. Feraday, G. H. Chalder, and K. D. Cotman, "Irradiation of U_3Si–Interim Results After 7360 MWd/tonne Irradiation," AECL-2648, Atomic Energy of Canada Ltd. (1967).

9. R. B. Mathews and M. L. Swanson, "Swelling of Uranium Silicide Fuel During Postirradiation Heating," *Nucl. Technol.*, **26**, 278 (1975).

10. C. R. Tipton, Ed., *Reactor Handbook, Vol. I, Materials*, 2nd ed., p. 201, Interscience Publishers, Inc., New York (1960).

11. H. Etherington, *Nuclear Engineering Handbook*, pp. 10-32, McGraw-Hill Book Company, New York (1958).

12. F. A. Rough, "An Evaluation of Data on Zr-U Alloys," BMI-1030, Battelle Memorial Institute (1955).

13. C. R. Tipton, Ed., *Reactor Handbook, Vol. I, Materials*, 2nd ed., pp. 175-204, Interscience Publishers, Inc., New York (1960).

14. S. Cerni, "SNAP-23A Phase I Quarterly Progress Report, September 1-November 30, 1967," WANL-3800-9, Westinghouse Astronuclear Laboratory (1967).

15. H. D. Weiss, "Properties of Mulberry," UCID-15170, University of California, Lawrence Livermore Laboratory (1967).

16. S. McLain and J. H. Martens, Eds., *Reactor Handbook, Vol. IV, Engineering*, 2nd ed., p. 177, Interscience Publishers, Inc., New York (1964).

17. G. B. Zorzoli, "Use of Metallic Thorium for LWBRs and LWRs," *Nucl. Technol.*, **20**, 109 (1973).

18. C. R. Tipton, Ed., *Reactor Handbook, Vol. I, Materials*, 2nd ed., pp. 315-321, Interscience Publishers, Inc., New York (1960).

19. C. E. Weber and H. Hirsch, "Dispersion Type Fuel Elements," *Proc. Int. Conf. Peaceful Uses Atomic Energy*, **9**, 196 (1956).

20. Max Jakob, *Heat Transfer*, Vol. I, p. 85, John Wiley & Sons, Inc., Publishers, New York (1949).

21. J. P. Stora, "Thermal Conductivity of Two-Phase Solid Bodies," *Nucl. Technol.*, 17, 225 (1973).

22. D. A. G. Bruggeman, *Anal. Phys.*, 24, 636 (1935).

23. H. Fricke, *Phys. Rev.*, 24, 575 (1924).

24. C. R. Tipton, Ed., *Reactor Handbook, Vol. I, Materials*, 2nd ed., p. 295, Interscience Publishers, Inc., New York (1960).

25. C. M. Friedrich and W. H. Guilinger, "CYGRO-2A, FORTRAN IV Computer Program for Stress Analysis of the Growth of Cylindrical Fuel Elements with Fission Gas Bubbles," WAPD-TM-547, Bettis Atomic Power Laboratory (1966).

26. D. R. De Hales and G. R. Horn, "Evolution of Uranium Dioxide Structure During Irradiation of Fuel Rods," *J. Nucl. Mater.*, 8, 20 (1963).

27. F. H. Nichols, "Theory of Columnar Grain Growth and Central Void Formation in Oxide Fuel Rods," *J. Nucl. Mater.*, 22, 214 (1967).

28. J. L. Bates and W. E. Roake, "Irradiation of Fuel Elements Containing Uranium Dioxide Powder," Reprint V-90-1959, Fifth Nucl. Energy Conf., Engineering Joint Council, New York (Apr. 1959).

29. J. D. Eichenberg, P. W. Frank, T. J. Kisiel, B. Lustman, and K. H. Vogel, "Effects of Irradiation on Bulk Uranium Dioxide," WAPD-183, Bettis Atomic Power Laboratory (1957).

30. M. F. Lyons, R. C. Nelson, T. J. Pashos, and B. Weidenbaum, "Uranium Dioxide Fuel Rod Operation with Gross Central Melting," *Trans. Am. Nucl. Soc.*, 6, 155 (1963).

31. C. Lepescky, G. M. Testa, H. Houggard, and K. W. Jones, "Experimental Investigation of In-Reactor Molten Fuel Performance," *Nucl. Technol.*, 16, 367 (1972).

32. J. D. Eichenberg, P. W. Frank, T. J. Kisiel, B. Lustman, and K. H. Vogel, "Effects of Irradiation on Bulk Uranium Dioxide," Fuel Elements Conf., Paris, France, TID-7546, U. S. Atomic Energy Commission (1958).

33. J. F. Kerrisk and D. G. Clifton, "Smoothed Values of the Enthalpy and Heat Capacity of UO_2," *Nucl. Technol.*, 16, 531 (1972).

34. A. W. Lemmon, "Analysis of Efficiency of Cooling and Extent of Reaction in the PWR," EADL-1, Battelle Northwest Laboratory report to Bettis Atomic Power Laboratory.

35. J. A. Christensen, "Thermal Expansion and Change in Volume in UO_2 in Melting," *J. Am. Ceram. Soc.*, 46, 607 (1963).

36. T. G. Godfrey, W. Fulkerson, T. G. Kollie, J. P. Moore, and D. L. McElroy, "Thermal Conductivity of Uranium Dioxide and Armco Iron by an Improved Radial Flow Technique," ORNL-3556, Oak Ridge National Laboratory (1964).

37. J. F. May, M. J. F. Notley, R. L. Stoute, and J. A. L. Robertson, "Observation on Thermal Conductivity of Uranium Dioxide," AECL-1641, Atomic Energy of Canada Ltd. (Nov. 1962).

38. J. Belle, R. M. Berman, W. F. Bourgeois, J. Cohen, and R. C. Daniel, "Thermal Conductivity of Bulk Oxide Fuels," WAPD-TM-586, Bettis Atomic Power Laboratory (Feb. 1967).

39. J. L. Bates, "Thermal Conductivity of Uranium Dioxide Improves at High Temperatures," *Nucleonics*, 19, 6, 83 (June 1961).

40. T. J. Nishijima, "Thermal Conductivity of Sintered Uranium Dioxide and Al_2O_3 at High Temperatures," *J. Am. Ceram. Soc.*, 48, 31 (1965).

41. A. D. Feith, "Thermal Conductivity of Uranium Dioxide by Radial Heat Flow Method," TM-63-9-5, General Electric Company (1963).

42. O. J. C. Runnals, "Uranium Dioxide Fuel Elements," CRL-55, National Research Council of Canada, Chalk River (1959).

43. M. F. Lyons, D. H. Coplin, T. J. Pashos, and B. Weidenbaum, "UO_2 Pellet Thermal Conductivity from Irradiations with Central Melting," *Trans. Am. Nucl. Soc.*, 7, 1, 106 (1964).

44. J. P. Stora, D. B. deDesigorger, R. Deimas, P. Deschamps, B. Travard, and C. Rionigot, "Conductibilité Thermique de l'Oxyde d'Uranium Fritté dans les Conditions d'Utilisation en Pile," CEA-R-2586, Commissariat à l'Energie Atomique (1964).

45. P. E. MacDonald and J. Weisman, "Effect of Pellet Cracking on Light Water Reactor Fuel Temperature," *Nucl. Technol.*, 31, 357 (1976).

46. J. A. Christensen, "Irradiation Effects on Uranium Dioxide Melting," HW-69234, Hanford Works, Richland, Washington (Mar. 1962).

47. P. E. MacDonald and I-Chih Wang, "Fuel Element Design," *Elements of Nuclear Reactor Design*, Chap. 10, J. Weisman, Ed., Elsevier Scientific Publishing Company, Amsterdam, The Netherlands (1977).

48. P. Guinet, H. Vaugoyeau, and P. Blum, "The UO_2-U System at High Temperatures," CEA-R-3060, Commissariat à l'Energie Atomique (1966).

49. J. Belle, Ed., *Uranium Dioxide Properties and Nuclear Applications*, U. S. Atomic Energy Commission, Washington, D. C. (1961).

50. P. E. MacDonald and L. B. Thompson, "MATPRO—Handbook of Materials Properties for Use in Analyses of Light Water Reactor Fuel Rod Behavior," ANCR-1263, Idaho Nuclear Engineering Laboratory (1976).

51. A. T. Jeffs, "Thermal Conductivity of ThO_2-PuO_2 During Irradiation," *Trans. Am. Nucl. Soc.*, 11, 497 (1968).

52. T. G. Godfrey, J. A. Woolley, and J. M. Leitmaker, "Thermodynamic Functions of Nuclear Materials," ORNL-TM-1596, Oak Ridge National Laboratory (1966).

53. L. F. Epstein, "Ideal Solution Behavior and Heats of Fusion from the UO_2-PuO_2 Phase Diagram," *J. Nucl. Mater.*, 23, 340 (1967).

54. J. L. Krankota and C. N. Craig, "Melting Point of High Burnup PuO_2-UO_2," *Trans. Am. Nucl. Soc.*, 11, 132 (1968).

55. J. F. Lagedrost, D. F. Askey, V. W. Storhok, and J. E. Gates, "Thermal Conductivity of PuO_2 as Determined from Thermal Diffusivity Measurements," *Nucl. Appl.*, 4, 54 (1968).

56. F. W. Albaugh, Battelle Northwest Laboratory Communication to *Reactor Materials*, 20, 207 (1968).

57. F. J. Hetzler, T. E. Lannin, K. J. Perry, and E. L. Zerbroski, "Thermal Conductivity of Uranium and Uranium Plutonium Oxide," GEAP-48799, General Electric Company (1967).

58. A. Biancheria et al., WARD-4135-1, Westinghouse Advanced Reactor Division (1969).

59. C. R. Tipton, Ed., *Reactor Handbook, Vol. I, Materials*, 2nd ed., p. 497, Interscience Publishers, Inc., New York (1960).

60. R. N. Duncan, W. H. Artl, and J. S. Atkinson, "Toward Low-Cost High Performance BWR Fuel," *Nucleonics*, **23**, *4*, 50 (1965).

61. E. F. Ibrahim, "An Equation for Creep of Cold Worked Zircaloy Pressure Tube Material," AECL-2928, Atomic Energy of Canada Ltd. (1965).

62. H. Stehle, W. Kaden, and R. Manzel, "External Corrosion of Cladding in PWRs," *Nucl. Eng. Des.*, **33**, 155 (1975).

63. W. R. Smalley, WCAP-3385-56, Westinghouse Electric Corporation (1971).

64. D. O. Pickman, "Internal Cladding Corrosion Effects," *Nucl. Eng. Des.*, **33**, 141 (1975).

65. R. D. Page, "Engineering Performance of Canada's UO_2 Fuel Assemblies for Heavy-Water Power Reactors," *Proc. IAEA Symp. Heavy-Water Power Reactors*, p. 749, Vienna (1967).

66. J. A. L. Robertson et al., AECL-4520, Atomic Energy of Canada Ltd. (1973).

67. R. J. Allio and K. C. Thomas, "Improved Iron-Base Cladding for Water Reactors," *Nucleonics*, **23**, *6*, 72 (1965).

68. C. R. Tipton, Ed., *Reactor Handbook, Vol. I, Materials*, 2nd ed., p. 312, Interscience Publishers, Inc., New York (1960).

69. J. H. Wright, "Supercritical Pressure Reactor Technology," Paper 65-8, Westinghouse Nuclear Power Seminar, Westinghouse Electric Corporation (1965).

70. H. Stehle, H. Assman, and F. Wunderlich, "Uranium Dioxide Properties for LWR Fuel Rods," *Nucl. Eng. Des.*, **33**, 230 (1975).

71. A. G. Evans and R. W. Davidge, *J. Nucl. Mater.*, **33**, 249 (1969).

72. R. F. Canon et al., "Steady State Creep Model for UO_2," *J. Am. Ceram. Soc.*, **54**, 105 (1971).

73. C. S. Olsen, "Steady State Creep Model for UO_2," *Trans. Am. Nucl. Soc.*, **22**, 210 (1975).

74. Y. H. Sun and D. Okrent, "A Simplified Method of Computing Clad and Fuel Strain and Stress During Irradiation," UCLA-Eng 7591, University of California at Los Angeles (1975).

75. H. Assman and H. Stehle, "Densification Effects in Sintered UO_2," *Proc. Reactortagung*, Karslruhe, Federal Republic of Germany (1973).

76. W. Chubb, A. C. Hott, B. M. Argall, and G. R. Kilp, "The Influence of Fuel Microstructures on In-Pile Densification," *Nucl. Technol.*, **26**, 486 (1975).

77. E. Rolstad, A. Hanevik, and K. D. Knudsen, "Measurements of the Length Changes of UO_2 Fuel Pellets During Irradiation," Paper 2/6, Enlarged Halden HPG Mtg., Sanjaford, Norway (1973).

78. J. Weisman, Ed., *Elements of Nuclear Reactor Design*, Elsevier Scientific Publishing Company, Amsterdam, The Netherlands (1977).

79. M. O. Marlowe, "In-Reactor Densification Behavior of UO_2," NEDO-12440, General Electric Company (1973).

80. R. L. Coble, "Sintering Crystalline Solids—I. Intermediate and Final State Diffusion Modes," *J. Appl. Phys.*, **32**, 787 (1961).

81. B. J. Buescher and G. R. Horn, "Densification Kinetics in UO_2 Fuels," *Trans. Am. Nucl. Soc.*, **22**, 205 (1975).

82. J. R. MacEwan and V. B. Lawson, "Grain Growth in Sintered Uranium Dioxide," *J. Am. Ceram. Soc.*, **45**, 42 (1962).

83. F. A. Nichols, "Pore Migration in Ceramic Fuel Elements," *J. Nucl. Mater.*, **27**, 137 (1968).

84. P. Huebotter, "Fast Reactor Fuel Element Design," paper presented to the 7th Ann. AMU-ANL Faculty Student Conf. (1968).

85. R. Godesar, M. Guyette, and N. Hoppe, "COMETHE-II—A Computer Code for Predicting the Mechanical and Thermal Behavior of a Fuel Pin," *Nucl. Technol.*, **9**, 205 (1970).

86. D. A. Collins and R. Hargreaves, "Performance Limiting Phenomena in Irradiated UO_2," *Proc. London Conf. on Nuclear Fuel Performance*, British Nuclear Energy Society, London (1973).

87. D. J. Clough, AERE-5948, U. K. Atomic Energy Research Establishment, Harwell, England.

88. F. A. Nichols and H. R. Warner, "A Statistical Fuel Swelling and Fission Gas Release Model," *Nucl. Technol.*, **9**, 148 (1970).

89. F. A. Nichols, H. R. Warner, H. Ocken, and S. H. Leiden, "Swelling and Gas Release Models for Oxide Fuels," *Trans. Am. Nucl. Soc., Suppl. 1*, **14**, 12 (1971).

90. Che Yu-Li, S. R. Pati, R. B. Poeppel, and R. O. Scattergood, "Some Considerations of the Behavior of Fission Gas Bubbles in Mixed Oxide Fuels," *Nucl. Technol.*, **9**, 188 (1970).

91. R. M. Carroll and O. Sissman, "Fission Gas Release During Fissioning of Uranium Dioxide," *Nucl. Appl.*, **2**, 142 (1966).

92. R. Soulhier, "Fission Gas Release from Uranium Dioxide During Irradiation Up to 2000°C," *Nucl. Appl.*, **2**, 138 (1966).

93. A. H. Booth, "A Method of Calculating Fission Gas Diffusion from UO_2 Fuel and Its Application to X-2 Loop Test," CRDC-721, National Research Council of Canada, Chalk River (1957).

94. C. Ronchi and H. Matzke, "Fuel and Fuel Elements for Fast Reactors," p. 57, International Atomic Energy Agency, Vienna (1974).

95. W. B. Lewis, "Engineering for the Fission Gas in Uranium Dioxide Fuel," *Nucl. Appl.*, **2**, 171 (1966).

96. J. F. Hoffman and D. H. Coplin, "Release of Fission Gases from Uranium Dioxide Fuel Operated at High Temperatures," GEAP-4596, General Electric Company (1964).

97. C. E. Beyer and C. R. Hann, "Prediction of Fission Gas Release from UO_2 Fuel," BNWL-1875, Battelle Northwest Laboratory (1973).

98. J. Weisman, P. E. MacDonald, A. I. Miller, and H. Ferrari, "Fission Gas Release from UO_2 Fuel Rods with Time Varying Power Histories," *Trans. Am. Nucl. Soc.*, **12**, 900 (1969).

99. E. E. Gruber, "Transient Fission-Gas Release from Oxide Fuels," *Trans. Am. Nucl. Soc.*, **22**, 418 (1975).

100. B. Lustman, "Fission Gas Pressure Within PWR Core Fuel Rods and Proposed PWR Core 2 Fuel Elements," WAPD-PWR-PMM-1034, Bettis Atomic Power Laboratory (Jan. 1957).

101. I. Cohen, B. Lustman, and J. D. Eichenberg, "Measurement of Thermal Conductivity of Metal Clad Uranium Dioxide Rods During Irradiation," WAPD-228, Bettis Atomic Power Laboratory (1960).

102. M. Knudsen, *Kinetic Theory of Gases*, John Wiley & Sons, Inc., Publishers, New York (1950).

103. R. A. Dean, "Thermal Contact Conductance," MS Thesis, University of Pittsburgh (1963).

104. D. D. Lanning and C. R. Hann, "Review of Methods Applicable to the Calculation of Gap Conductance in Zircaloy-Clad UO_2 Fuel Rods," BNWL-1894, Battelle Northwest Laboratory (1975).

105. A. Ullman, R. Acharya, and D. R. Olander, "Thermal Accommodation of Inert Gases on Stainless Steel and UO_2," *J. Nucl. Mater.*, **51**, 277 (1974).

106. H. Von Ubisch, S. Hall, and R. Srivastav, "Thermal Conductivities of Mixture of Fission Product Gases with Helium and Argon," Paper F/143, 2nd Geneva Conf. Peaceful Uses of Atomic Energy, International Atomic Energy Agency, Vienna (1957).

107. J. N. Cetinkale and M. Fishenden, "Thermal Conductance of Metal Surfaces in Contact," *Proc. General Discussion on Heat Transfer*, p. 271, Institute of Mechanical Engineers, London (Sep. 1951).

108. H. Fenech and W. M. Rohsenow, "A Prediction of Thermal Conduction of Metallic Surfaces in Contact," *J. Heat Transfer*, **85**, 15 (1963).

109. R. G. Wheeler, "Thermal Contact Conductance of Fuel Element Materials," HW-60343, Hanford Works, Richland, Washington (1959).

110. A. C. Rapier, T. M. Jones, and J. E. McIntosch, "The Thermal Conductance of Uranium Dioxide/Stainless Steel Interfaces," *Int. J. Heat Mass Transfer*, **6**, 397 (1963).

111. A. M. Ross and R. L. Stoute, "Heat Transfer Coefficient Between Uranium Dioxide and Zircaloy-2," CFRD-1075, Atomic Energy of Canada Ltd. (1962).

112. F. R. Campbell and R. DesHaies, "Effect of Gas Pressure on Fuel/Sheath Heat Transfer," *Trans. Am. Nucl. Soc.*, **21**, 380 (1975).

113. R. H. Veckerman and R. Harris, "Thermal Conductivity and Temperature Jump Distance of Gas Mixtures," *Trans. Am. Nucl. Soc.*, **22**, 523 (1975).

114. Y. L. Shlykov, *Teploenergetika*, **12**, *10*, 79 (1965); see also, English translation, "Thermal Contact Resistance," *Therm. Eng.*, **12**, 102 (Apr. 1966).

115. I. L. Shvets and E. P. Dyban, *Inzh. Fiz. Zh.*, 3 (1964).

116. N. Todreas and G. Jacobs, "Thermal Contact Conductance of Reactor Fuel Elements," *Nucl. Sci. Eng.*, **50**, 283 (1973).

117. A. Calza-Bini, G. Cosali, G. Filaccheoni, M. Lanehi, A. Nobili, E. Pesce, U. Rocca, and P. L. Rotaloni, "In-Pile Measurement of Fuel Cladding Conductance for Pelleted and Vipac Zircaloy-2 Sheathed Fuel Pins," *Nucl. Technol.*, **25**, 103 (1975).

118. D. D. Lanning, C. R. Hann, and E. S. Gilbert, "Statistical Analysis and Modeling of Gap Conductance Data for Reactor Fuel Rods Containing UO_2 Pellets," BNWL-1832, Battelle Northwest Laboratory (1974).

119. G. Kjaerheim and E. Rolstaad, "In-Pile Determination of UO_2 Thermal Conductivity, Density Effects, and Gap Conductance," HPR-80, Halden, Norway (1967).

120. R. L. Mehan and P. W. Weisinger, "Mechanical Properties of Zircaloy-2," KAPL-2110, Knolls Atomic Power Laboratory (1961).

121. C. C. Busby, "Longitudinal Uniaxial Tensile and Compressive Properties," in *Properties of Zircaloy-4 Tubing*, C. R. Woods, Ed., WAPD-TM-585, Bettis Atomic Power Laboratory (1966).

122. J. E. Irvin, "Effect of Irradiation and Environment on the Mechanical Properties of Hydrogen Pickup of Zircaloy," in *Zirconium and Its Alloys*, Electrochemical Society (1966).

123. E. F. Ibrahim, "In-Reactor Creep of Zircaloy-2 at 260 to 300°C," *J. Nucl. Mater.*, **45**, 169 (1973).

124. P. E. MacDonald and L. B. Thompson, "MATPRO—Handbook of Materials Properties for Use in Analyses of Light Water Reactor Fuel Rod Behavior," ANCR-1263, Idaho Nuclear Engineering Laboratory (1976).

125. D. L. Hagrman, "Cladding Axial Growth," *Quarterly Report on Reactor Safety Programs Sponsored by the Nuclear Regulatory Commission, July-Sept. 1975*, Sec. 4.1, ANCR-1296, U. S. Nuclear Regulatory Commission, Washington, D. C. (1975).

126. J. R. Fagan and J. O. Mingle, "Effect of Axial Heat Conduction in Fuel Plates on Maximum Heat Flow Rates and Temperatures," *Nucl. Sci. Eng.*, **18**, 443 (1964).

127. J. A. L. Robertson, " $\int kdt$ in Fuel Irradiations," CRFD-835, Atomic Energy of Canada Ltd. (1959).

128. M. F. Lyons, D. H. Coplin, H. Hausner, B. Weidenbaum, and T. J. Pashos, "Uranium Dioxide Powder and Pellet Thermal Conductivity During Irradiation," GEAP-5100-1, General Electric Company (1966).

129. D. J. Clough and J. B. Sayers, "Measurement of Thermal Conductivity of Uranium Dioxide Under Irradiation in the Temperature Range 150-1000°C," AERE-R-4690, U. K. Atomic Energy Research Establishment, Harwell, England (1964).

130. R. Soulhier, "Comportment des Gas de Fission Dans les Elements Combustibles à Oxide d'Uranium—Tentative de Synthese des Resultats Experimentaux," DM/1629, Commissariat à l'Energie Atomique (1967).

131. J. Randles, "Heat Diffusion in Cylindrical Fuel Elements of Water Cooled Reactors," AEEW-R96, U. K. Atomic Energy Establishment, Winfrith, England (1961).

132. D. G. Andrews and M. D. Dixmier, "Calculation of Temperature Distributions in Fuel Rods with Varying Conductivity and Asymmetric Flux Distribution," *Nucl. Sci. Eng.*, **36**, 259 (1969).

133. D. E. Lamkin and R. E. Brehm, "Analytical Stress Analysis Solution for a Simplified Model of a Reactor Fuel Element," *Nucl. Technol.*, **27**, 273 (1975).

134. C. M. Friedrich and H. W. Guilinger, "CYGRO-2, A FORTRAN IV Computer Program for Stress Analysis and the Growth of Cylindrical Fuel Elements with Fission Gas Bubbles," WAPD-TM-547, Bettis Atomic Power Laboratory (1966).

135. C. K. Cheng and B. M. Ma, "Thermal, Radiation, and Mechanical Analysis of Unsteady State Fuel Restructuring of Cylindrical Oxide Elements in Fast Reactors," *Nucl. Sci. Eng.*, **48**, 138 (1972).

136. M. F. Lyons, D. H. Coplin, and B. Weidenbaum, "Analyses of UO_2 Grain Growth Data from Out-of-Pile Experiments," GEAP-4411, General Electric Company (1963).

137. C. W. Sayles, "A Three Region Analytical Model for the Description of the Thermal Behavior of Low Density Oxide Fuel Rods in a Fast Reactor Environment," *Trans. Am. Nucl. Soc.*, **10**, 458 (1967).

138. E. Duncombe, C. M. Friedrich, and W. H. Guilinger, "An Analytic Model for the Prediction of In-Pile Behavior of Oxide Fuel Rods," *Nucl. Technol.*, **12**, 194 (1971).

139. V. Z. Jankus and R. W. Weeks, "LIFE II—A Computer Analysis of Fast Reactor Fuel Element Behavior as a Function of Reactor Operating History," *Nucl. Eng. Des.*, **18**, *1*, 83 (1972).

140. J. Rest, "SST: A Computer Code to Predict Fuel Response and Fission Release from Light Water Reactor Fuels During Steady-State and Transient Conditions," *Trans. Am. Nucl. Soc.*, **22**, 462 (1975).

141. J. A. Dearien et al., "FRAP-S: A Computer Code for the Steady State Design of Oxide Fuel Rods, Vol. I—Analytical Models and Input," I-309-13.1, Aerojet Nuclear Company (1975).

142. C. R. Hann, C. E. Beyer, and L. J. Parchen, "GAPCON-THERMAL-1: A Computer Program for Calculating the Gap Conductance of Oxide Fuel Pins," BNWL-1778, Battelle Northwest Laboratory (1973).

143. C. A. Olson and L. H. Boman, "Three-Dimensional Thermal Effects in PWR Fuel Rods," WA/HT-78, presented at the Winter ASME Mtg., Houston, Texas, American Society of Mechanical Engineers, New York (1975).

144. B. L. Pierce, "TAPA—A Program for Computing Transient or Steady State Temperature Distribution," WANL-TME-1872, Westinghouse Electric Corporation (1969).

145. F. E. Tippets, "An Analysis of the Transient Conduction of Heat in a Long Cylindrical Fuel Element for Nuclear Reactors," HW-41896, Hanford Works, Richland, Washington (1956).

146. L. A. Matsch, "Transient One-Dimensional Temperature Distribution in a Two Region Infinite Circular Cylinder," MS Thesis, University of Pittsburgh (1960).

147. D. S. Rowe, "COBRA-IIIC: A Digital Computer Program for Steady State and Transient Thermal Hydraulic Analysis of Rod Bundle Nuclear Fuel Elements," BNWL-1695, Pacific Northwest Laboratory (1973).

148. C. F. Bonilla, "Analysis of a Power Failure Incident for the Preliminary Conceptual Design of a Small Size Pressurized Water Reactor," Columbia University Report to Gibbs and Hill, Inc. (1959).

149. L. S. Tong, "Simplified Calculation of Thermal Transient of a Uranium Dioxide Fuel Rod," *Nucl. Sci. Eng.*, 11, 340 (1961).

150. W. L. Chen, M. Ishii, and M. A. Grolmes, "Simple Heat Conductance Model with Phase Change for a Reactor Fuel Pin," *Nucl. Sci. Eng.*, 60, 452 (1976).

151. A. Pearson and C. G. Lennox, *The Technology of Nuclear Reactor Safety*, T. J. Thompson and J. G. Beckerley, Eds., Vol. I, pp. 399-407, MIT Press, Cambridge, Massachusetts (1964).

152. L. Baker and L. C. Just, "Studies of Metal-Water Reactions at High Temperatures," *Experimental and Theoretical Studies of the Zirconium-Water Reaction*, ANL-6548, Argonne National Laboratory (1962).

153. L. N. Grossman and D. M. Rooney, "Interfacial Reaction Between UO_2 and Zircaloy-2," GEAP-4679, General Electric Company (1965).

154. D. O. Hobson and P. L. Rittenhouse, "Embrittlement of Zircaloy Clad Fuel Rods by Steam During LOCA Transients," ORNL-4758, Oak Ridge National Laboratory (1972).

155. D. O. Hobson, "Ductile-Brittle Behavior of Zircaloy Fuel Cladding," CONF-730304, U. S. Atomic Energy Commission, Washington, D. C. (1973).

156. "Acceptance Criteria for Emergency Core Cooling Systems, Light Water-Cooled Nuclear Power Reactors," Docket No. RM-50-1, U. S. Atomic Energy Commission, Washington, D. C. (1974).

157. D. O. Pickman, "Interaction Between Fuel Pins and Assembly Components," *Nucl. Eng. Des.*, 33, 125 (1975).

158. J. A. Dearien, L. J. Stefkin, M. P. Bohn, R. C. Young, and R. L. Benedetti, "FRAP-T2: A Computer Code for the Transient Analysis of Oxide Fuel Rods, Vol. I—Analytical Models and Input," I-309-3-53, Aerojet Nuclear Company (1975).

159. K. Hannerz and G. Verterlund, "Zircaloy Cladding Mechanical Properties," *Nucl. Eng. Des.*, 33, 205 (1975).

160. D. O. Hobson and P. L. Rittenhouse, "Deformation and Rupture Behavior of Light-Water Reactor Fuel Cladding," ORNL-4727, Oak Ridge National Laboratory (1971).

161. F. F. Cadek et al., "PWR Flecht—Full Length Emergency Core Cooling Heat Transfer—Final Report," WCAP-7665, Westinghouse Electric Corporation (1971).

162. P. C. Hall and R. B. Duffy, "A Method of Calculating the Effect of Clad Ballooning on Loss-of-Coolant-Accident Temperature Transients," *Nucl. Sci. Eng.*, 58, 1 (1975).

163. E. Rolstad, "Model for Prediction of Fuel Failures," *Trans. Am. Nucl. Soc.*, 24, 163 (1976).

164. J. C. Wood and B. A. Surette, "Pellet/Cladding Interaction—Evaluation of Lubrication by Graphite," *Trans. Am. Nucl. Soc.*, 28, 208 (1978).

3

HYDRODYNAMICS

The rate of heat removal from a reactor core and the dynamic forces on the core and internals depend strongly on the flow behavior of the system. In this chapter, the basic flow characteristics of friction, turbulence, void distribution, and depressurization are described. Specific applications such as hydrodynamic problems in the reactor vessel and flow instability in a core are then discussed.

3-1 BASIC FLOW CHARACTERISTICS

Under steady-state conditions, the total pressure drop across a flow loop, ΔP_{total}, is the sum of the frictional, acceleration, and elevation pressure drops.

$$\Delta P_{total} = \Delta P_{friction} + \Delta P_{acceleration} + \Delta P_{elev} \ . \qquad (3.1)$$

However, in single-phase flow, the integrals of the acceleration and elevation changes around a loop are nearly zero (exactly zero under isothermal conditions). Determination of the total pressure loss is then reduced to the determination of the frictional pressure loss. The pressure loss must, of course, be matched by the pressure rise across the pump. Hence, flow through the loop is fixed at the intersection of the head-flow curve of the pump and the flow-head loss curve of the loop.

3-1.1 Single-Phase Flow Friction

Single-phase flow pressure drop is generally calculated in accordance with

$$\Delta P_{friction} = \left(\frac{fL}{D_e}\right)\frac{\rho V^2}{2g_c} + \sum_i \left(K_i \frac{\rho V_i^2}{2g_c}\right) , \qquad (3.2)$$

where

f = skin friction factor due to wall shear in a straight flow channel. Empirical values for f can be obtained from the Moody curve[1] in Fig. 3-1.

K_i = form friction factor due to the i'th change in the cross section or restriction in the flow channel. The value of K for sudden expansion K_e and

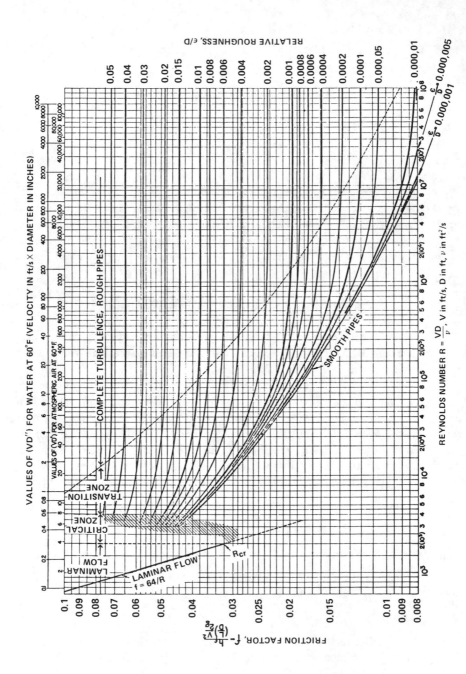

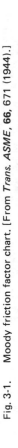

Fig. 3-1. Moody friction factor chart. [From *Trans. ASME*, **66**, 671 (1944).]

for sudden contraction K_c are given by Kays and London.[2] Their values at high Reynolds numbers are reproduced in Fig. 3-2.

V = fluid velocity through the channel

V_i = fluid velocity in the smaller channel at area change i

D_e = equivalent diameter of the straight flow channel, 4 X cross-sectional area/perimeter.

Analytical expressions have been developed to describe the friction characteristics of various flow regimes. For circular channels, the laminar friction factor can be calculated

$$f = 64/\text{Re} , \qquad (3.3)$$

where Re is the Reynolds number. The turbulent friction factor in a smooth channel can be calculated from the approximate relationship

$$f = C\,\text{Re}^{-0.2} , \qquad (3.4)$$

where C equals 0.184 for isothermal flow in commercially available smooth tubes. The effect of surface roughness is indicated in Fig. 3-1 in terms of the ratio of the height of the surface projections, ϵ, to pipe diameter, D. For commercial steel pipe, ϵ can be taken as $\sim$0.00015 ft.

3-1.1.1 Bends, Valves, Fittings, and Restrictions

Pressure losses across bends, valves, and fittings are expressed in either terms of a loss coefficient, K_l, or a length of pipe, L_e, having an equivalent pressure drop. Thus, pressure loss ΔP_l across a given component can be calculated by

$$\Delta P_l = K_l \frac{\rho V^2}{2g_c} = \left(\frac{fL_e}{D_e}\right)\frac{\rho V^2}{2g_c} . \qquad (3.5)$$

The reader is referred to Refs. 3, 4, and 5 for tabulations of K_l and (L_e/D_e) for commonly encountered components.

Pressure loss across a long restriction in a flow path can be estimated by adding frictional loss to expansion and contraction losses estimated via the K_i values of Fig. 3-2. However, when the restriction is so short (e.g., a perforated plate) that the vena contracta occurs beyond the restriction, then the pressure drop should be obtained from

$$\Delta P_l = \frac{G^2}{2g_c\sigma^2\rho C^2}(1 - 2\sigma C + \sigma^2 C^2) , \qquad (3.6)$$

where

G = mass velocity upstream of restriction

C = vena contracta area ratio

σ = ratio of restricted flow area to full flow area.

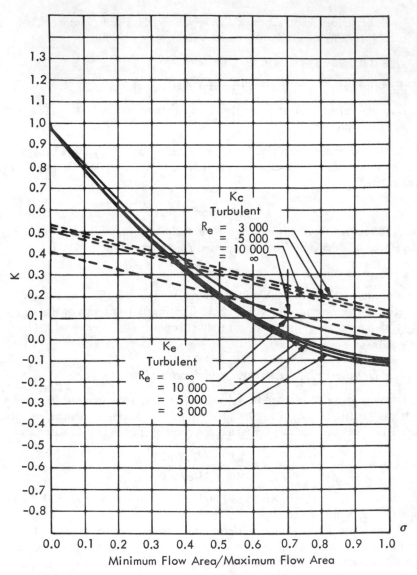

Fig. 3-2.   Form friction factors for sudden expansions and contractions. [From W. M. Kays and A. L. London, *Compact Heat Exchangers,* The National Press, Palo Alto, California (1955).]

3-1.1.2 *Effect of Heating*

In a channel where one wall is heated or cooled, the friction factor is changed due to the change of viscosity near the wall. For turbulent flow, Seider and Tate[6] suggested

$$f/f_{ISO} = \left(\frac{\mu_{wall}}{\mu_{bulk}}\right)^{0.14} , \tag{3.7}$$

where f_{ISO} is the friction factor measured under isothermal conditions, and μ is fluid viscosity. For water flowing at 2000 psia, the effect has been correlated by Esselman et al.,[7] LeTourneau et al.,[8] and Mendler et al.[9] as

$$f/f_{ISO} = 1 - 0.0019 \, \Delta T_f , \tag{3.8}$$

where

$$\Delta T_f = q''/h$$

$$h = 0.023 \, (k_b/D_e)(Re_b)^{0.8}(Pr_b)^{0.4}$$

k_b = thermal conductivity of fluid at the bulk temperature

Pr_b, Re_b = Prandtl and Reynolds numbers, respectively, at bulk temperature.

A limiting value of 0.85 has been given for Eq. (3.8) by its authors.

Rohsenow and Clark[10] examined the Sieder and Tate[6] relationship using water flowing at pressures of 1500 to 2000 psi. They found the exponent for Eq. (3.7) was 0.60 for the frictional pressure loss alone, but 0.14 for the pressure loss due to both momentum and frictional losses. They noted that many of the data points gave values for f/f_{ISO} below 0.85. The relationship

$$\frac{f_{friction}}{f_{ISO\text{-}friction}} = \left(\frac{\mu_{wall}}{\mu_{bulk}}\right)^{0.6} \tag{3.9}$$

is recommended.

3-1.1.3 Core Pressure Drop

The reactor cores of almost all pressurized water reactors consist of an array of rod bundles. Hence, flow characteristics within the core are those of the bundles of which it is composed. Core pressure drop can be considered to be composed of three components:

1. pressure loss along the bundle of bare rods

2. pressure loss across the spacer grids or wires

3. pressure loss at core entrance and exit.

Entrance and exit losses can be evaluated as those due to sudden expansions and contractions, as previously described.

The pressure loss along bare rods is usually predicted by assuming similarity to that for flow inside a tube and by using the equivalent diameter concept. The validity of using the D_e concept in predicting pressure drop in a rod bundle was investigated by Deissler and Taylor.[11] By an iterative procedure, they plotted the velocity profile in square and triangular lattices as shown in Figs. 3-3 and 3-4, respectively. These velocity profiles were calculated from a generalized velocity

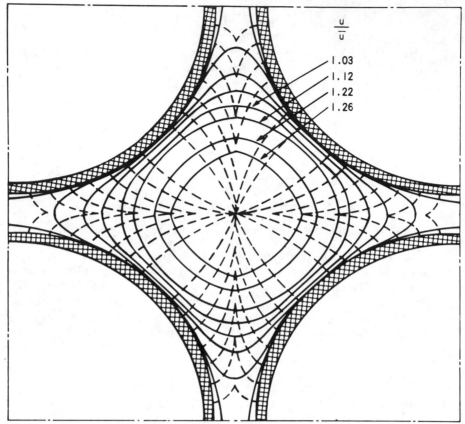

Square array; Reynolds number, 40 000; spacing parameter, 1.1.

Fig. 3-3.   Predicted velocity distribution between tubes (from Ref. 11).

profile[12] based on Deissler[13] and Laufer's[14] data. From these velocity profiles, friction factors and Reynolds numbers can be calculated by integrating the profiles to obtain bulk (or average) velocities, since both friction factors and Reynolds numbers are based on bulk velocities. Predicted friction factors for triangular and square arrays are shown in Figs. 3-5 and 3-6. Deissler and Taylor's[11] predictions are in the vicinity of the line based on circular tube data. However, there is an effect of pitch/diameter (P/D) that would not be observed if the D_e concept strictly held. The experimental data shown have too much scatter to indicate any trend.

A number of other investigators have considered the effect of the P/D ratio: Miller et al.[15] tested a 37-rod bundle of 3-ft heated length with 0.625-in.-o.d. rods in a triangular lattice of 1.46 p/d ratio and found the friction factor represented by

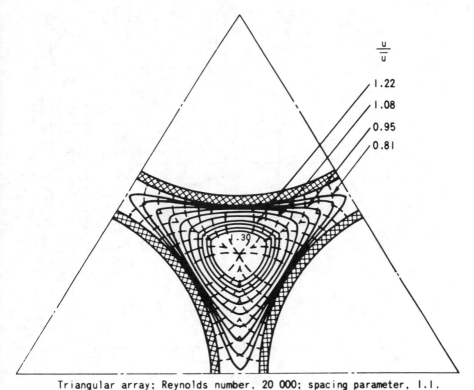

$$\frac{u}{\bar{u}}$$

1.22

1.08

0.95

0.81

Triangular array; Reynolds number, 20 000; spacing parameter, 1.1.

Fig. 3-4.   Predicted velocity distribution between tubes (from Ref. 11).

$$f = 0.296 \ Re^{-0.2} \ , \tag{3.10}$$

which is 60% higher than obtained by fitting the Moody curve.

Le Tourneau et al.[8] tested rod bundles of square lattice with P/D ratios of 1.12 and 1.20 and of triangular lattice with a P/D ratio of 1.12. These data fall within a band between the Moody curve for smooth tubes and a curve 10% below that in a Reynolds number range of 3×10^3 to 10^5.

Wantland[16] tested a 100-rod bundle of square lattice of $\frac{3}{16}$-in. o.d. and 6-ft length with a P/D of 1.106 and 102-rod bundle of triangular lattice of $\frac{3}{16}$-in. rod o.d. and 6-ft length with a P/D ratio of 1.190. For Prandtl numbers between 3 and 6, he obtained

1. For square lattices at $Re = 10^3 - 10^4$ and $Pr = 3 - 6$,

$$f = 1.76 \ Re^{-0.39} \ , \tag{3.11}$$

which is ~30% higher than Eq. (3.4).

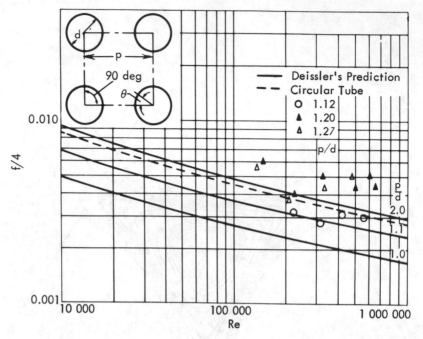

Fig. 3-5.   Comparison of predicted and measured friction factors for square lattices (from Ref. 11).

2. For triangular lattices at Re = $2 \times 10^3 - 1 \times 10^4$,

$$f = 90.0 \, Re^{-1} + 0.0082 \ , \tag{3.12}$$

which is ~25% higher than Eq. (3.4).

Dingee et al.[17] tested 9-rod bundles in the water of 0.5-in. rod o.d., 24-in.-long square and triangular lattices with a P/D of 1.12, 1.20, and 1.27. Square lattice data are 5 to 20% higher than Deissler and Taylor's theoretical prediction.[18] The triangular lattice data agree with the same prediction, but with a scatter of about ±15%.

More recently, Trupp and Azad[19] measured velocity distributions, eddy diffusivities, and friction factors with air flow in triangular array bundles. These data indicated friction factors somewhat higher than Deissler and Taylor's predictions.[11] For Reynolds numbers between 10^4 and 10^5, their data at P/D = 1.2 were ~17% higher than circular tube data. The data at P/D = 1.5 were ~27% higher than circular tube data. Trupp and Azad[19] correlated their triangular array data by

$$f = C \, Re^{-n} \ , \tag{3.13}$$

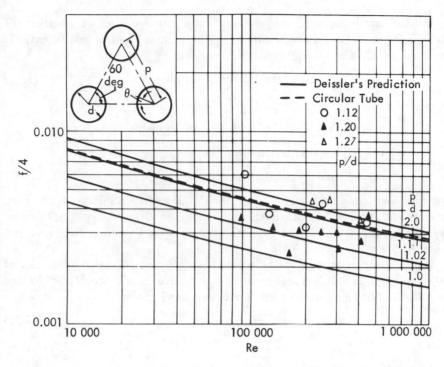

Fig. 3-6.   Comparison of predicted and measured friction factors for triangular lattices (from Ref. 11).

where

$$C = 0.287\,[(2\,\sqrt{3}/\pi)(P/D)^2 - 1.30] \left.\begin{array}{l} \\ \\ \end{array}\right\} \quad \begin{array}{l} 10^4 \leqslant Re \leqslant 10^5 \\ \\ 1.2 \leqslant P/D \leqslant 1.5 \ . \end{array}$$
$$n = 0.368\,(P/D)^{-1.358}$$

Pressure losses across spacer grids or wires are from drag-type pressure losses. These are often calculated using pressure-loss coefficients for sudden contractions and expansions. Berringer and Bishop[20] tested the Yankee fuel rod bundle with both cylindrical ferrule spacers and egg-crate strap grids. The test section bundle was composed of 36 rods of 0.335-in. o.d. in a square lattice with a P/D of 1.27. The ferrule is 0.5 in. long with a 0.265-in. o.d. and a 20-mil wall thickness. The strap is $1\frac{1}{8}$ in. high and $\frac{1}{32}$ in. thick.

Comparison of the ferrule- and strap-pressure drops indicates that pressure drop across a level of straps was more than four times that across a full-ferruled section. Prediction of the pressure drop across the ferrule or strap using the loss coefficients for sudden contraction and expansion agrees with the data within 17%. Friction factors for the bare rod section are ~15% higher than the Moody curve in a pipe in the Reynolds number range of 10 000 to 30 000.

De Stordeur[21] measured the pressure drop characteristics of a variety of spacers and grids. He correlated his results in terms of a drag coefficient, C_s. Pressure drop ΔP_s across the grid or spacer was given by

$$\Delta P_s = \rho C_s V_s^2 s / 2 g_c A \ , \tag{3.14}$$

where

A = unrestricted flow area away from the grid or spacer

V_s = velocity in the spacer region

s = projected frontal area of the spacer.

The drag coefficient used was a function of the Reynolds number for a given spacer or grid type. At high Reynolds number (Re $\simeq 10^5$) strap-type grids showed drag coefficients of ~1.65. The pressure drop across a crossed circular wire grid was ~10% lower.

On the basis of tests of several grids, Rehme[22] concluded that the effect of the ratio (s/A) is more pronounced than indicated by de Stordeur.[21] Rehme concludes that grid pressure drop data are better correlated by

$$\Delta P_s = C_V (\rho V_B^2 / 2 g_c)(s/A)^2 \ , \tag{3.15}$$

where

C_V = modified drag coefficient

V_B = average bundle fluid velocity.

Drag coefficient C_V is a function of the average bundle Reynold's number. Rehme's data indicate that for square arrays, $C_V = 9.5$ at Re $= 10^4$ and $C_V = 6.5$ at Re $= 10^5$.

Rehme[22] also correlated a wide selection of data on the total pressure losses in wire-wrapped bundles. These pressure losses depend on the P/D ratio, helical spacer wire pitch H', and the number of rods. Rehme correlated rod-bundle pressure loss data in terms of a modified form of the friction factor, f'. Pressure loss is given by

$$\Delta P = f' \left(\frac{S_b}{S_t} \right) \frac{\rho V_{eff}^2 L}{2 g_c D_e} \ , \tag{3.16}$$

where

S_b/S_t = ratio of the wetted perimeter of rods and wires to the total wetted perimeter

V_{eff} = effective velocity accounting for swirl flow due to wire wraps

$\quad = (V_B) \{ (P/D)^{6.5} + [7.6 (d_m/H')(P/D)^2]^{2.16} \}^{1/2}$

d_m = mean diameter of wire wraps

f' = modified friction factor $= (64/N_{Re}') + 0.0816/(N_{Re}')^{0.133}$

N'_{Re} = modified Reynolds number = $\rho V_{eff} D_e / \mu$.

The total pressure drop across a core or reactor vessel is often expressed as

$$\Delta P_{core} = K_{core} \frac{\rho V^2}{2g_c}$$

$$\Delta P_{vessel} = K_{vessel} \frac{\rho V^2}{2g_c} \ , \tag{3.17}$$

where ρ is average density in the core and V is average flow velocity inside the core. For example, K values for the Yankee Rowe reactor are

$$K_{core} = 15 \ , \qquad K_{vessel} = 34 \ ,$$

and K values for the SELNI reactor are

$$K_{core} = 18 \ , \qquad K_{vessel} = 52 \ .$$

Lateral flow (normal to the fuel rod axis) occurs in a core where there are significant pressure differences between assemblies. In the lateral direction, the core appears as a large tube bank. Idelchik[5] correlates turbulent flow friction factors for this situation by equations of the form

$$f_L = A \ Re_L^{-0.2} \ , \tag{3.18}$$

where A is a function of rod pitch and diameter (Ref. 5) and Re_L is the lateral Reynolds number based on rod diameter.

3-1.2 Two-Phase Flow Void Fraction and Pressure Drop

Although there is no net void at the outlet of a PWR under normal operating conditions, there is substantial subcooled boiling within the cores of modern plants. Bulk boiling is allowed in the hottest channels; further, during most accident situations, boiling occurs throughout most of the core. Neither steady-state nor transient behavior can be understood without a knowledge of two-phase flow fundamentals. Since the details of two-phase flow behavior were discussed in Ref. 23, only a summary is presented here.

3-1.2.1 Void Fraction

Because of the relatively low density of steam compared to liquid water, the steam vapor phase is called void. Void fraction, α, is defined as the ratio of local vapor volume to the total local flow volume, that is,

$$\alpha = \frac{A_v}{A_v + A_l} \ , \tag{3.19}$$

where A_l is the cross-sectional area occupied by the liquid and A_v is the cross-sectional area occupied by the vapor. The average density, $\bar{\rho}$, of a two-phase mixture then becomes

$$\bar{\rho} = \alpha\rho_v + (1 - \alpha)\rho_l \ , \tag{3.20}$$

where ρ_v and ρ_l are the vapor and liquid densities, respectively. Slip ratio S is defined as the ratio of vapor velocity to liquid velocity

$$S = \frac{V_v}{V_l} = \frac{\chi}{1 - \chi} \frac{1 - \alpha}{\alpha} \frac{\rho_l}{\rho_v} \ , \tag{3.21}$$

where χ = flow quality = mass flow rate of vapor/mass flow rate of the mixture.

Void fraction in a heated channel can be described with the aid of Fig. 3-7. Initially, the entire fluid is highly subcooled and no void is present. When the liquid adjacent to the wall becomes superheated, bubbles form. In Region I, where the bulk of the liquid is still highly subcooled, the voids are a thin layer attached to the wall. In Region II, where subcooling is low, void fraction increases and bubbles are detached from the wall and recondense very slowly. In Region III, the bulk liquid has reached saturation and bulk boiling occurs.

Maurer[24] correlated void fraction in Region I in terms of δ, the effective thickness of the vapor film on the wall. Thus,

$$\alpha = \frac{4\delta}{D_e} \tag{3.22}$$

and δ is the maximum of 0.00033 ft or

$$\delta = \frac{q_b'' k_l \mathrm{Pr}_l}{1.07 \, h^2 (T_{\mathrm{sat}} - T_{\mathrm{bulk}})} \ , \tag{3.23}$$

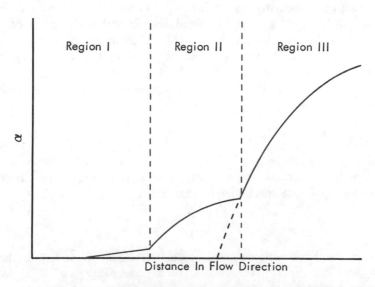

Fig. 3-7.   Void fraction in boiling region (from Ref. 25).

where

$q_b'' =$ heat transfer rate due to boiling $= q_{total}'' - h(T_{sat} - T_b)$

$h =$ forced convection heat transfer coefficient

$k_l =$ thermal conductivity of the liquid

$\text{Pr}_l =$ Prandtl number of the liquid.

Bowring[25] proposed that the detached region begins at a critical fluid subcooling that depends on flow velocity and heat flux. His criterion can be written as

$$(T_s - T_0)V_{in}/q'' = 14 + 0.1 P \ , \tag{3.24}$$

where

$V_{in} =$ inlet velocity, cm/s

$q'' = W/cm^2$

$P =$ pressure, atm

$T_0 =$ fluid temperature at detachment point, °C

$T_s =$ saturation temperature, °C.

Maitra and Raju[26] compared the Bowring criterion to their own experimental measurements and obtained reasonable agreement. However, they did not find good agreement between Bowring's criterion and data in the literature.

Griffith et al.[27] suggested an alternative criterion for the detachment point which has also been used for design purposes. Staub[28] and Levy[29] developed criteria that depend on heat flux, velocity, and dimensionless thickness of the bubble layer.

Bowring[25] also suggested an empirical method for predicting α in Region II. He assumed a slip ratio of unity and no appreciable condensation of the vapor. However, this latter assumption may not always be valid. Levy[29] proposed a relatively simple scheme for estimating the true vapor weight fraction in Region II. The pseudo quality so calculated is then substituted in an appropriate relationship between α and X.

Thom et al.[30] correlated void fractions in subcooled boiling for the 750- to 1000-psi range. Their correlation, which is in terms of apparent density $\bar{\rho}$, appears somewhat simpler. Tong[31] extended their correlation to 2000 psi; the modified correlation is

$$\bar{\rho} = \rho_l - \frac{\gamma X'(\rho_l - \rho_v)}{1 + X'(\gamma' - 1)} \ , \tag{3.25}$$

where

ρ_l, ρ_v = liquid and vapor densities, respectively

$$X' = \left(H_{\text{in}} - \frac{4Lq''}{D_e G} - \beta H_l\right) \Big/ (H_v - \beta H_l)$$

$$\gamma' = \rho_l/(\rho_v S) = \exp\{4.216\,[(y - 8.353)^2/(8.353)^2 - 1]^{1/2}\}$$

$$y = \ln(P/3206)$$

P = pressure in psia

β = ratio of enthalpy at inception of detached region to the saturation enthalpy = $1 - 0.15\,q''/G$

H_{in} = inlet enthalpy to channel, Btu/lb

H_v, H_l = saturation enthalpies of vapor and liquid, respectively, Btu/lb

G = mass velocity, lb/(h ft^2)

L = heated length, ft.

The above equation was developed from uniform heat flux data; although it has been applied to channels with nonuniform heat flux and mixing by replacing $(4L_{q''}/D_e G)$ by the enthalpy rise to the point in question, experimental verification is lacking.

When bulk boiling begins, quality X is determined from fluid enthalpy H by

$$H = XH_v + (1 - X)H_l \ . \tag{3.26}$$

Void fraction is obtained from the quality. However, void fraction and slip ratio during bulk boiling vary with the flow pattern. At the pressures and flows encountered in a PWR, the pattern existing at the onset of boiling will be small vapor bubbles dispersed through a liquid core (bubbly flow). At higher void fractions, small bubbles coalesce into large elongated bubbles, and slugs of liquid and vapor follow each other (slug flow). If enthalpy is increased beyond the range normally observed in a PWR, vapor slugs form a continuous core with an annulus of liquid adjacent to the wall (annular flow).

The simplest treatment of bubbly flow is Bankoff's.[32] By assuming a power law distribution for velocity and void fraction, the following equation relating x and α was derived:

$$1/X = 1 - \rho_l/\rho_v(1 - C/\alpha) \ , \tag{3.27}$$

with the flow factor C defined by

$$C = \alpha + (1 - \alpha)/S \ . \tag{3.28}$$

Bankoff[32] proposed that constant C can be obtained from

$$C = 0.71 + 0.0001\,P \ , \tag{3.29}$$

where P is in psia. Revised values for C were proposed by Kholodovski[33] and Hughmark.[34] By correlating available data, Hughmark concluded that C was a function of parameter Z, where

$$Z = \left[\frac{D_e G}{\mu_l(1 - \alpha) + \mu_g \alpha}\right]^{1/6} \left(\frac{G^2}{\rho_h^2 g D_e}\right)^{1/8} \left[\frac{(1 - X)\rho_g + X\rho_l}{(1 - X)\rho_g}\right]^{1/4}, \qquad (3.30)$$

where

ρ_h = mixture density obtained by assuming no slip

μ_l, μ_g = liquid and vapor viscosities, respectively.

All other symbols have their previous meaning. The relationship between C and Z proposed by Hughmark is given in Table 3-I. Use of Hughmark's relationship leads to slip ratios that approach 1.0 as mass velocity increases. This is in accord with experimental observations. The Hughmark correlation tends to give poor results for alphas (α) greater than ~0.8. At high α, it is probably preferable to obtain the relationship between α and X by simply extrapolating the Hughmark X versus α curve developed for $\alpha < 0.75$ so that the curve goes through $\alpha = 1$ at X = 1.0.

More recently, Zuber and Findlay[35] proposed a more sophisticated model applicable to bubbly and slug flow regimes. They define local superficial velocities $u = (Q_v + Q_l)/A$, $u_v = (Q_v/A)$, and $u_l = (Q_l/A)$ based on total flow area A. Velocity and void distributions were assumed to follow

$$\frac{u'}{u_c} = 1 - \left(\frac{r}{R}\right)^n, \qquad (3.31)$$

$$\frac{\alpha - \alpha_w}{\alpha_c - \alpha_w} = 1 - \left(\frac{r}{R}\right)^n, \qquad (3.32)$$

where r is radial distance, R is tube radius, and subscripts c and w refer to centerline and wall conditions, respectively. For adiabatic flow, α_w is taken as zero, and C, the flow factor, is given by

$$C = \frac{\alpha}{\beta} = \frac{\bar{u}_v}{\alpha(\bar{u})} = \frac{n+1}{n+2}, \qquad (3.33)$$

TABLE 3-I

RELATIONSHIP BETWEEN FLOW FACTOR C AND PARAMETER Z

Z	C	Z	C	Z	C	Z	C	Z	C
1.3	0.185	3.0	0.49	6.0	0.72	15	0.808	70	0.93
1.5	0.225	4.0	0.605	8.0	0.767	20	0.83	130	0.98
2.0	0.325	5.0	0.675	10	0.78	40	0.88		

where n is a constant and $\bar{u}_v$ and $\bar{u}$ are the mean values of u_v and u, respectively. To estimate bubble slip velocity, Zuber and Findlay use terminal bubble velocities estimated by Griffith and Wallis[36] and Harmathy.[37] For the slug flow regime, they obtain V_v, the weighted mean velocity of the vapor, as

$$\bar{V}_v = \left(\frac{\bar{u}_v}{\alpha}\right) = \frac{(\bar{u})}{C} + 0.35 \left(\frac{g\Delta\rho D}{\rho_l}\right)^{1/2} , \tag{3.34}$$

and for the bubbly flow regime

$$\bar{V}_v = \left(\frac{\bar{u}_v}{\alpha}\right) = \frac{(\bar{u})}{C} + 1.53 \left(\frac{\sigma g g_c \Delta\rho}{\rho_l^2}\right)^{1/4} , \tag{3.35}$$

where $\Delta\rho$ equals $\rho_l - \rho_v$ and σ equals surface tension.

For small terminal bubble rise velocities, where

$$C = \bar{u}/\bar{V}_v , \tag{3.36}$$

the value of C is less than unity when $\alpha_c > \alpha_w$, and greater than unity when $\alpha < \alpha_w$. For the former case, C is close to 0.8.

Other descriptions of the bubbly and slug flow regimes were suggested by Marchaterre and Hoglund[38] and Petrick.[39]

The most widely used void fraction correlation in the annular flow regime is that of Martinelli and Nelson,[40] who presented the data as a plot of α versus quality for a series of different pressures. Thom[41] modified these on the basis of improved data and proposed a correlation of the form

$$\alpha = \frac{\gamma_1 X}{1 + X(\gamma_1 - 1)} , \tag{3.37}$$

where X is flow quality and γ_1 is an empirical constant equal to $P/(\rho_l S)$. Thom proposed the values of γ_1 shown in Table 3-II.

3-1.2.2 Pressure Drop in Pipes and Rod Bundles

In the highly subcooled boiling region where only attached wall voidage is present, it has sometimes been assumed that single-phase pressure drop calculations are adequate. However, this is not the case at high heat fluxes typical of reactor cores. The measured pressure drop is significantly above the single-phase value.

TABLE 3-II

PARAMETER VALUES FOR THE THOM VOID FRACTION CORRELATION

Values of γ							
Pressure (psia)	14.7	250	600	1250	2100	3000	3206
γ	246	40.0	20.0	9.80	4.95	2.15	1.0

Correlations were proposed by Owens and Shrock,[42] Le Tourneau and Troy,[43] and Tarasova et al.[44] The correlation of Tarasova et al.,[44] which includes both friction and acceleration losses, is particularly convenient. Tarasova gives the ratio of boiling pressure drop, $(dP/dz)_b$, to isothermal (nonboiling) pressure drop, $(dP/dz)_0$, as

$$(dP/dz)_b/(dP/dz)_0 = 1 + [20N/(1.315 - N)] [q''v_g/(H_{fg}V_{in})]^{0.7}(v_g/v_l)^{0.08} , \qquad (3.38)$$

where

$$N = (H - H_0)/(H_{sat} - H_0)$$

q'' = wall heat flux, Btu/h ft^2

v_g, v_l = specific volume of gas and liquid, respectively, ft^3/lb

H = liquid enthalpy, Btu/lb

H_{sat}, H_0 = liquid enthalpy at saturation and inception of boiling, respectively, Btu/lb

V_{in} = inlet velocity, ft/h

H_{fg} = enthalpy change on evaporation, Btu/lb.

The foregoing correlation was developed for data with heat fluxes between 0.18 to 0.55×10^6 Btu/h ft^2, pressures between 142 and 2840 psi, and G between 1.0 and 2.2×10^6 lb/h ft^2.

In the detached and bulk boiling regions, the increase in fluid volume significantly increases the velocity of the stream and, therefore, its momentum; consequently, all three pressure drop components indicated by Eq. (3.1)—frictional loss, momentum change, and elevation pressure drop—need to be considered separately under these conditions. At the low void fractions ($\alpha < 0.30$) and high mass flows usually encountered in a water-cooled reactor, two-phase pressure drop can be evaluated from the homogeneous model proposed by Owens.[45] The elevation pressure drop is given by

$$\Delta P_{elev} = \int_0^L \rho dL \approx \bar{\rho}\Delta L . \qquad (3.39)$$

For a local boiling flow, $\bar{\rho}$ can be obtained from Eq. (3.25). The friction pressure drop is obtained from

$$\Delta P_{friction} = \frac{f\Delta L}{D_e} \frac{\bar{v}G^2}{2g_c} , \qquad (3.40)$$

and the acceleration pressure drop term is obtained from

$$\Delta P_{acc} = \frac{G^2}{g_c} \frac{d\bar{v}}{dL} \Delta L , \qquad (3.41)$$

where G is the mass velocity (lb/h ft^2) and $\bar{v}$ is the average specific volume of the mixture at the location considered.

Evaluation of friction factor f in Eq. (3.40) requires estimating a Reynolds number. Hence, viscosity $\bar{\mu}$ of the two-phase mixture is needed. The formulation suggested by McAdams et al.[46] has generally been used.

$$\bar{\mu} = \chi\mu_g + (1 - \chi)\mu_l ,\qquad(3.42)$$

where

μ_g, μ_l = viscosities of gas and liquid, respectively

$\bar{\mu}$ = viscosity of two-phase mixture.

However, Choe[47] obtained improved agreement with available data by assuming that the viscosity of a bubbly mixture is actually higher than that of a pure liquid. Choe[47] suggests that fluid viscosity can be obtained from the following:

$$\bar{\mu} = \mu_l \exp[2.5/(1 - 39\alpha/64)] \qquad \alpha \leqslant 0.5 ,$$

or

$$\bar{\mu} = \mu_g + (\mu_c - \mu_g)\left(\frac{1}{\alpha} - 1\right)^3 \qquad \alpha > 0.5 ,\qquad(3.43)$$

where

$$\mu_c = \bar{\mu} \quad \text{at} \quad \alpha = 0.5 .$$

The pressure drop obtained during high void-fraction flow again depends on the flow pattern; however, a general empirical correlation for calculating the two-phase frictional pressure drop was developed by Baroczy.[48] His work is considered an extension of that of Lockhart and Martinelli[49] and Martinelli and Nelson,[40] all of whom defined the two-phase pressure drop in terms of a single-phase pressure drop. Baroczy writes

$$\phi_{lo}^2 = \left(\frac{\Delta P}{\Delta L}\right)_{TP}\bigg/\left(\frac{\Delta P}{\Delta L}\right)_l ,\qquad(3.44)$$

where

ϕ_{lo}^2 = two-phase frictional pressure drop multiplier

$(\Delta P/\Delta L)_{TP}$ = two-phase pressure drop per unit length

$(\Delta P/\Delta L)_l$ = single-phase pressure drop obtained at the same mass velocity when the fluid is entirely liquid.

By defining a property index $[(\mu_l/\mu_v)^{0.2}(\rho_l/\rho_v)]$, Baroczy[48] obtained a correlation for ϕ_{lo}^2 that was independent of pressure. He also observed that his correlation could be used with the gas-phase pressure drop $(\Delta P/\Delta L)_g$ by noting

$$\frac{(\Delta P/\Delta L)_l}{(\Delta P/\Delta L)_g} = \frac{(\mu_l/\mu_v)^{0.2}}{(\rho_l/\rho_v)} .\qquad(3.45)$$

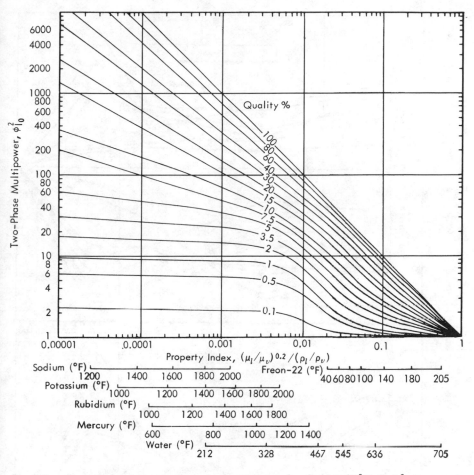

Fig. 3-8. Two-phase friction pressure drop correlation for $G = 1 \times 10^6$ lb/(h ft^2) (from Ref. 48).

His correlation is given in two sets of curves:

1. a plot of the two-phase multiplier ϕ_{lo}^2 as a function of property index $(\mu_l/\mu_v)^{0.2}/(\rho_l/\rho_v)$, as shown in Fig. 3-8

2. plots of a two-phase multiplier ratio as a function of property index, quality, and mass velocity as shown in Fig. 3-9.

The latter ratio multiplies ϕ_{lo}^2 whenever G deviates from 1×10^6 lb/(h ft^2).

At high void fractions, use of the homogeneous model for computation of the acceleration pressure drop is not desirable since the effect of slip can be significant. It is convenient to write the acceleration pressure drop as

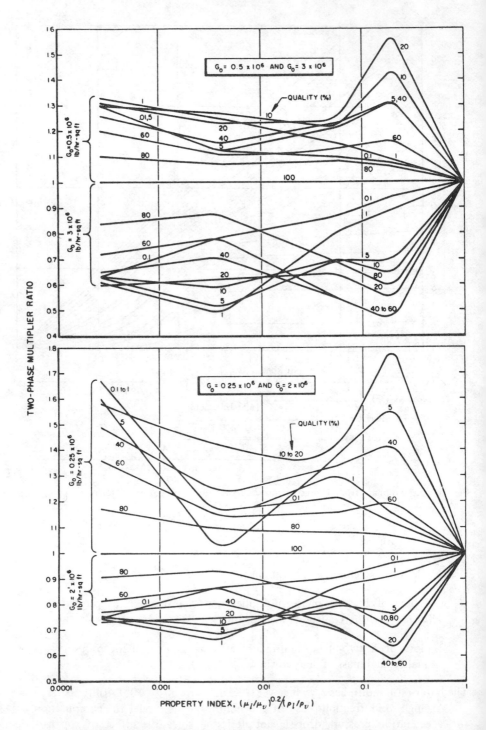

Fig. 3-9.   Mass velocity correction versus property index (from Ref. 48).

$$\Delta P_{acc} = r \, \frac{G^2}{g_c \rho_l} \, , \tag{3.46}$$

where r is a dimensionless acceleration multiplier. When the inlet fluid contains no voids, r is given by

$$r = \frac{(1 - \chi_e)^2}{(1 - \alpha_e)} + \frac{\chi_e^2}{\alpha_e} \left(\frac{\rho_l}{\rho_v} \right)_{sat} - 1 \, , \tag{3.47}$$

where

χ_e = flow quality at channel exit

α_e = void fraction at channel exit.

By proper evaluation of α, the effect of slip is included. The value of r is independent of the manner of heat addition (sinusoidal, uniform, etc.) along the channel.

In bulk boiling, the effect of wall heat flux on friction factors has often been ignored. However, Tarasova et al.[44] recommend that the adiabatic multipliers be corrected by

$$\phi_{heat}^2 / \phi_{adiabatic}^2 = 1 + 0.99 (q''/G)^{0.7} \, , \tag{3.48}$$

where q'' is given in Btu/h ft^2 and G is expressed in lb/h ft^2 (for $710 < P < 2840$ psia, $0.37 \times 10^6 < G < 1.9 \times 10^6$, and $0.032 \times 10^6 < q'' < 0.53 \times 10^6$).

3-1.2.3 Pressure Drop Across Area Changes, Restrictions, and Fittings

Pressure differences across area changes can be computed by one-dimensional momentum balances. For an abrupt expansion, Lottes[50] reports that the total pressure difference can be obtained from

$$\Delta P_e = \frac{G_T^2 \sigma^2}{\rho_l g_c} \left(\left[\frac{\rho_l}{\rho_g} \chi^2 \left(\frac{1}{\alpha_1 \sigma} - \frac{1}{\alpha_2} \right) \right] + \left\{ (1 - \chi)^2 \left[\frac{1}{\sigma(1 - \alpha_1)} - \frac{1}{(1 - \alpha_2)} \right] \right\} \right) \, , \tag{3.49}$$

where

α_1, α_2 = void fractions upstream and downstream, respectively, of expansion

σ = area ratio

G = total mass flow rate, mass/time area

ρ_l, ρ_g = liquid and gas densities, respectively

χ = mixture quality.

Janssen[51] gives the pressure loss across a contraction as

$$\Delta P_c = \frac{G_l^2}{2g_c\rho_l} \left(\frac{1}{\sigma^2} \left\{ \frac{\rho_l}{\rho_g} \chi^2 \bar{\alpha}_1 \left(\frac{1}{C^2\alpha_3^2} - \frac{1}{\alpha_2^2} \right) + (1-\chi)^2(1-\bar{\alpha}_1) \left[\frac{1}{C^2(1-\alpha_3)^2} - \frac{1}{(1-\alpha_2)^2} \right] \right\} \right.$$

$$- \frac{2}{\sigma^2} \left\{ \frac{\rho_l}{\rho_g} x^2 \left(\frac{1}{C\alpha_3} - \frac{1}{\alpha_2} \right) + (1-x)^2 \left[\frac{1}{C(1-\alpha_3)} - \frac{1}{(1-\alpha_2)} \right] \right\}$$

$$\left. + \frac{\rho_l}{\rho_g} x^2\bar{\alpha}_2 \left(\frac{1}{\sigma^2\alpha_2^2} - \frac{1}{\alpha_1^2} \right) + (1-x)^2(1-\bar{\alpha}_2) \left[\frac{1}{\sigma^2(1-\alpha_2)^2} - \frac{1}{(1-\alpha_1)^2} \right] \right) , \quad (3.50)$$

where

C = vena contracta area ratio

α_1, α_2 = void fractions upstream and downstream, respectively, of contraction

$\bar{\alpha}_1 = (\alpha_3 + \alpha_2)/2$

$\bar{\alpha}_2 = (\alpha_1 + \alpha_2)/2$

α_3 = void fraction at vena contracta.

On the basis of an extensive analysis of available data, Weisman et al.[52] concluded that for abrupt expansions, α_1 and α_2 should be evaluated by assuming slip flow. They recommend Hughmark's[34] relationship for obtaining α from χ. For abrupt contractions, they again recommended that α_1 and α_2 be obtained by assuming slip flow and Hughmark's correlation.[34] However, substantial mixing occurs at the vena contracta and

$$\alpha_3 = \alpha_{slip} + (\alpha_{homog} - \alpha_{slip})K' , \quad (3.51)$$

where

$\alpha_{slip} = \alpha$, which would be computed for vena contracta size pipe in the absence of contraction

$K' = 1$ if $\alpha_2 \leqslant 0.5$

$K' = 1.5 - \alpha_2$ if $\alpha_2 > 0.5$.

It was also found that assuming homogeneous flow everywhere provided nearly as good a correlation of the data as Eq. (3.50) did. Then, total pressure drop across a contraction can be approximated by

$$\Delta P_c = \frac{G^2}{2\sigma^2 g_c} \left[\left(\frac{1}{C} - 1 \right)^2 + (1-\sigma)^2 \right] \left[\frac{\chi}{\rho_g} - \frac{(1-\chi)}{\rho_l} \right] . \quad (3.52)$$

Although this equation does not represent what is actually occurring, it is a useful tool.

When contraction and expansion are well separated, the total pressure loss is

$$\Delta P_t = \Delta P_c - \Delta P_e , \quad (3.53)$$

where ΔP_c and ΔP_e are obtained from Eqs. (3.49) and (3.50) and α_3 is obtained from Eq. (3.51). However, for restrictions with a length-to-diameter (L/D) ratio of <2, only partial mixing at the vena contracta is obtained. For these conditions, α_3 can be estimated from Eq. (3.51) with mixing factor K' calculated from

$$K' = C\,(\text{L/D})^b \,, \tag{3.54}$$

where

$$C = 1.48 - 1.57\alpha$$

$$b = 1.5\alpha - 0.34.$$

If the foregoing leads to a value of K' above 1 or below zero, K' is set equal to 1 or zero.

When the vena contracta is outside (beyond) the restriction, then a revised momentum balance must be used and we obtain

$$\Delta P_t = \frac{G_f^2 v_l}{2g_c\sigma^2}\frac{1}{C^2}\left\{\frac{\rho_l}{\rho_g}\,\chi^2\hat{\alpha}_3\left(\frac{1}{\alpha_3^2} - \frac{\sigma^2 C^2}{\alpha_2^2}\right) + (1-\chi)^2(1-\hat{\alpha}_3)\left[\frac{1}{(1-\alpha_3)^2} - \frac{\sigma^2 C^2}{(1-\alpha_2)^2}\right]\right.$$

$$\left. - 2\sigma C\left[\frac{\rho_l}{\rho_g}\,\chi^2\left(\frac{1}{\alpha_3} - \frac{\sigma C}{\alpha_2}\right) + (1-\chi)^2\left(\frac{1}{1-\alpha_3} - \frac{\sigma C}{1-\alpha_2}\right)\right]\right\}, \tag{3.55}$$

where

α_1, α_2 = void fractions upstream and downstream, respectively, of restriction

α_3 = void fraction at vena contracta

C = vena contracta area ratio

σ = ratio of restricted flow area to full flow area.

For short restrictions (L/D $\leqslant 0.5$), there appears to be little mixing at the vena contracta and α can be estimated, assuming slip flow; e.g., from Hughmark's correlation.[34]

At L/D ratios above 0.5, some mixing at the vena contracta is obtained and K', the mixing factor at the vena contracta, can be estimated from Eq. (3.54) with $C = -1.8\alpha + 1.05$ and $b = 3.1\,(\alpha - 0.36)$. Again, if this leads to a value for K' above 1 or below zero, K' is set at 1 or zero.

Only limited data on the two-phase pressure drop across bends and fittings have been reported. The best available correlation appears to be Chisholm's.[53] He suggests that the two-phase pressure drop, ΔP_{TP}, can be obtained from the single-phase pressure drop via

$$(\Delta P_{\text{TP}})/(\Delta P_l) = 1 + C(\Delta P_g/\Delta P_l)^{1/2} + (\Delta P_g/\Delta P_l) \tag{3.56}$$

where

$$C = [1 - (C_2 - 1)(v_{fg}/v_g)^{1/2}][(v_g/v_l)^{1/2} + (v_l/v_g)^{1/2}] \,,$$

and

$\Delta P_g, \Delta P_l$ = pressure drop with liquid or gas phase flowing alone

v_g, v_l = specific volumes of liquid and gas phases, respectively

C_2 = constant depending on ratio of bend radius R to pipe diameter D.

Table 3-III contains the values of C_2 which Chisholm found to fit available data on gradual bends. For 90-deg bends, it is suggested that $C_2 = 1 + 35N$, where N is the number of diameters of straight pipe with the same single-phase pressure drop as the bend.

Chisholm and Sutherland[54] recommend that Eq. (3.56) be used for calculating two-phase pressure loss across tees. They recommend that C_2 be set at 1.75.

3-1.3 Flow Mixing and Redistribution

In a rod bundle, flow channels formed by four adjacent fuel rods are open to each other through the gap between two neighboring fuel rods. The flow in one channel mixes with that of the others. Because of the pressure differential between the channels, there is also flow redistribution between channels. Local fluid mixing, caused by turbulence, reduces the enthalpy rise of a hot channel. On the other hand, flow leaving the hot channel increases its enthalpy rise. Calculating the net result is quite complicated, although the equation describing the enthalpy exchange between two open channels, m and n, can easily be written. The energy balance across axial increment ΔZ due to cross flow and mixing is

$$W_m \frac{\Delta H_m}{\Delta Z} = W_{n\text{-}m}(H_n - H_m) + \rho \epsilon (H_n - H_m) + q'' p_m \ , \tag{3.57}$$

where

p_m = heated perimeter of channel m, ft

W_m = axial flow rate in channel m, lb/s

$W_{n\text{-}m}$ = lateral cross-flow rate per unit length, lb/s ft

H_m, H_n = local enthalpy of channels m and n, respectively

TABLE 3-III
RELATIONSHIP BETWEEN C_2 AND R/D

C_2	R/D			
	1	3	5	7
Normal bend	4.35	3.4	2.2	1.0
Bend with upstream disturbance within 50 L/D	3.1	2.5	1.75	1.0

ΔH_m = enthalpy rise in channel m through ΔZ, Btu/lb

ϵ = mixing coefficient (a function of axial velocity and the gap between rods)

$\rho\epsilon$ = flow exchange rate per unit length, lb/s ft, or ~0.06 lb/s ft for a typical PWR fuel assembly without mixing enhancement devices.

3-1.3.1 Mixing

A number of studies have been directed toward evaluating the degree of mixing in rod bundles. Weisman et al.[55] used Rowe's[56] analytical studies of small size bundles. Since these studies concluded that cross-flow resistance was low, the analysis of the behavior of a 25-rod bundle by Weisman et al.[55] assumed that cross-flow resistance was so low that no lateral pressure existed across the bundle at any elevation. The interchannel flow so obtained allowed the enthalpy transfer due to diversion cross flow to be calculated. The enthalpy transfer due to mixing was taken as the difference between the total transfer and that due to cross flow. The mixing data were correlated by defining a thermal diffusion coefficient, α, such that

$$\alpha = \epsilon/Vb \ , \qquad (3.58)$$

where V is the superficial velocity (ft/s) and b is the gap between two rods. For their bundle (0.422-in.-diam rods on a 0.535-in.^{2} pitch), they found $\alpha \approx 0.076$ and essentially independent of mass velocity (G up to 2.75×10^6) and quality. Since their spacer grids contained small mixing vanes, it is expected that α for bundles without these vanes would be lower.

Some investigators preferred to correlate their data on the basis of a modified Peclet number, Pe, where

$$\mathrm{Pe} = VD_e/\epsilon \ , \qquad (3.59)$$

where D_e is equivalent diameter. Observe that the Peclet number equals $(D_e/\alpha b)$. For bundles of bare rods, Bell and LeTourneau[57] reported

$$1/\mathrm{Pe} = \epsilon/(VD_e) = 0.003 \ ; \qquad 10^4 < \mathrm{Re} < 5 \times 10^4 \ . \qquad (3.60)$$

They believed the data indicated that the correlation would be valid up to $\mathrm{Re} = 2 \times 10^5$. For rectangular channels, Sembler[58] reports

$$\epsilon/(VD_e) = 0.0015 \ ; \qquad \mathrm{Re} > 10^5 \ . \qquad (3.61)$$

Waters[59] reported on the mixing obtained in wire-wrapped bundles. He found that α increased from 0.1 to 1 as the wrap pitch changed from 0.7 to 3 wraps/ft.

Other investigators represented the degree of mixing in terms of w', the turbulent transverse fluctuating flow rate per foot of axial length [lb/(h ft)], where

$$w' = \rho e(l/D_e) \ , \qquad (3.62)$$

and

ρ = fluid density, lb/ft^3

e = eddy diffusivity, ft^2/h

l = Prandtl mixing length, ft

D_e = equivalent diameter of channel, ft.

Rowe and Angle[60] correlated their data by

$$w' = cGD_e\text{Re}^{-0.1} \ , \tag{3.63}$$

where

G = fluid mass velocity, $lb/(h \ ft^2)$

Re = Reynolds number

c = constant = 0.0062.

Rodgers and Rosehart[61] analyzed the data of a number of experimenters and found they could use Eq. (3.63) to represent the data if c = 0.004.

More recently, Rodgers and Rosehart[62] modified their correlation. For square pitch rods (adjacent channels with the same D_e), they propose

$$w' = 0.005bG\left(\frac{D_e}{D}\right)\text{Re}^{-0.1}\left(\frac{b}{D}\right)^{-0.894} \ , \tag{3.64}$$

where

G = mass velocity = m/A, $lb/(h \ ft^2)$

b = gap between rods, ft

D = rod diameter, ft.

This latter correlation indicates a slight effect of rod spacing in contrast to Eq. (3.63), which shows no such effect. The more recent work of Petrunik[63] and Singh[64] helps to elucidate this question. For P/D ratios in excess of 1.05, there is only a slight effect of gap spacing. At slightly lower ratios, they saw an increase in mixing with rod gap. Petrunik[63] believes there must be a region where there is a sharp decrease in w' since w' must go to zero as $b \to 0$. These conclusions are in agreement with the theoretical analysis of Van der Ros and Bogaart[65] and the results of Galbraith and Knudson.[66]

For the range of P/D ratios used in reactor design, it appears that the effect of gap spacing on turbulent mixing is small. Equation (3.61) or (3.64) appears to be adequate for computing in single-phase flow.

Mixing two-phase fluids under reactor conditions is still imperfectly understood. Modeling two-phase mixing in most computer codes used for rod bundle analysis is based on homogeneous theory. This can be visualized as the exchange of two packets of fluid between adjacent subchannels. In the equi-mass model, the packets are of equal mass and there is no net transfer of mass, only of heat and momentum. Enthalpy interchange due to mixing can be written as

$$w' (H_i - H_j) \ , \qquad (3.65)$$

where

H_i, H_j = mean enthalpy of channels i and j, respectively

w' = mass interchange between subchannels, lb/(h ft).

In the equi-volume model, the packets are of equal volume. When densities in the two subchannels are different, the mixing process transfers mass as well as heat and momentum. Hence, cross flow is not the only mechanism transferring mass and its definition and magnitude are changed accordingly.

Fundamental data on two-phase mixing under reactor conditions are not available. The most nearly representative data are those of Rowe and Angle.[67] They used lithium, deuterium, and tritium tracers to measure steam-water mixing in the presence of boiling at 750 and 400 psia. Their data differ substantially from single-phase observations. For void fractions below ~65 to 70%, mixing was enhanced as the void fraction increased. Overall mixing enhancement was greatest at the lowest velocity and largest gap spacing. The maximum mixing observed was ~10 times that observed in single-phase operation. Results with and without spacers were similar.

Mixing coefficients for low-pressure air-water systems were measured by Rudzinski et al.,[68] Gonzalez-Santalo and Griffith,[69] and du Bousquet and Bouré.[70] All these investigators saw trends similar to those observed by Rowe and Angle.[67]

At high void fractions, all investigators report an inverse relationship between quality (or void fraction) and mixing rate. Rudzinski et al.[68] observed that their peak mixing, as well as Rowe and Angle's,[67] occurred prior to the slug-annular flow transition. Singh and St. Pierre's[71] investigation mixing in the annular flow region confirms the inverse relationship with void fraction in that region.

Beus[72] devised a correlation that predicts the general form of the observed two-phase mixing data. He assumes that in the bubbly-slug flow regime, turbulent mixing is dominated by the churning motion of slugs and bubbles. In this regime, w'_I, the exchange flow per unit length, is given by

$$w'_I = w'_L + B_1 \left(\frac{AG}{D_e}\right)\left(\frac{\rho_l}{\rho_g}\right)\left(\frac{S-1}{S}\right)\chi \ , \qquad (3.66)$$

where

w'_L = exchange flow per unit length for pure liquid

A = channel flow area

S = slip ratio

χ = quality

ρ_l, ρ_g = liquid and vapor densities, respectively

B_1 = empirical constant.

Peak mixing, w_c', is assumed to take place at the slug-annular transition at quality X_c, given by the transition relationship as

$$X_c = \frac{\dfrac{0.4}{G}\,[g\rho_l D_e(\rho_l - \rho_g)]^{1/2} + 0.6}{0.6 + \left(\dfrac{\rho_l}{\rho_g}\right)^{1/2}} \ . \tag{3.67}$$

The peak mixing rate is determined by solving Eq. (3.66) when $X = X_c$.

The transition from peak mixing to mixing, w_g', obtained in the presence of pure vapor, is given by an empirical equation that has the form

$$w_{II}' = w_g' + (w_c' - w_g)\left(\frac{1 - K}{\dfrac{X}{X_c} - K}\right) , \tag{3.68}$$

where X is the empirical coefficient. Beus found that he could reasonably fit the available air-water and the moderate pressure steam-water mixing data by setting

$$B = 0.04\left(\frac{b}{D_e}\right)^{1.5} , \tag{3.69}$$

and

$$k = 0.57\ \mathrm{Re}^{0.0417} \ . \tag{3.70}$$

In view of the effect of void fraction on mixing, most reactor computations use empirical mixing parameters. While mixing parameters can be calculated from Eqs. (3.66) and (3.68), it is not certain whether these equations properly account for pressure effects. Mixing coefficients are more generally estimated by fitting coefficients to data from rod bundle heat transfer tests.

It is usually assumed that a single mixing parameter can be used for vapor and liquid. This is reasonable only if the mixing quality (quality of w') equals subchannel quality. Values of mixing quality were reported by Singh and St. Pierre.[71] For gap spacings of 40 mils and above, the mixing quality was close to the subchannel quality. At the lowest gap spacing, the mixing quality fell significantly below the subchannel quality. Since this effect occurs at a gap spacing considerably below that of reactor interest, the use of a single mixing coefficient for liquid and vapor does not appear unreasonable.

Under some conditions, it has not been possible to model the results of rod bundle tests satisfactorily using eddy diffusivity mixing alone. In some of their rod bundle tests, Lahey and Schraub[73] found that the quality in a particular corner channel was lower than that of the center channel, even when there was corner flux peaking. They proposed superposition of a void drift on eddy diffusivity mixing, with the drift proportional to the mass velocity gradient. By considering turbulent mixing to be on a volume-to-volume basis, they obtained m'', the net mass flux from channel i to j,

$$m'' = \frac{e(\rho_i - \rho_j)}{l} + e_{VD}\frac{(G_i - G_j)}{l} , \qquad (3.71)$$

where

m'' = transverse mass flux due to mixing and void drift, lb/(h ft^2)

l = characteristic mixing length, ft

e = eddy diffusivity, ft^2/h

e_{VD} = void drift diffusivity coefficient, ft

ρ_i, ρ_j = average density in channels i and j, respectively

G_i, G_j = average mass velocities in channels i and j, respectively, lb/(h ft^2).

A somewhat similar point of view has been put forward by Gonzalez-Santalo and Griffith.[69]

The formulation of Lahey and Schraub[73] and Gonzalez-Santalo and Griffith[69] are both approximate methods for dealing with annular flow. Annular flow occurs at void fractions in excess of those normally encountered in PWRs. Consideration of void drift effects is usually not needed in PWR modeling.

3-1.3.2 Flow Redistribution

Variation in hydraulic conditions among the various subchannels leads to differing axial pressure drops. Hence, at any given axial level there will be pressure gradients leading to lateral flow. In early subchannel computations, it was simply assumed that lateral flow velocity u could be obtained from

$$\Delta P = \frac{K_1 \rho u |u|}{2g_c} , \qquad (3.72)$$

where

ΔP = pressure difference between two adjacent assemblies

K_1 = lateral frictional resistance

ρ = fluid density.

However, it was recognized that the above formulation was inadequate since the inertial effect, due to high axial velocity, was neglected. Chelemer et al.[74] recognized the effect of axial velocity and, on the basis of single-phase data for flow through slots, computed K_1 from

$$(K_1/K_\infty - 1)K_1 = \gamma'(u/V)^{-2} , \qquad (3.73)$$

where

γ' = constant

V = axial velocity

K_∞ = value of K_1 as (V/u) approaches zero.

More recently, Kim and Stoudt[75] used a similar approach to correlate cross flow between bundles containing spacer grids. They proposed the average interbundle cross-flow velocity, u, could be correlated by an equation of the form

$$u = (2K_sV_1 + B) - [(2K_sV_1 + B)^2 - 4K_sC]^{1/2}/(2K_s\Delta xA_c/A_x) , \qquad (3.74)$$

where

$$B = 2(V_1 + V_2) + \Delta x(f_1V_1 + f_2V_2)/D$$

$$C = K_sV_1^2 - 2g\Delta P/\rho + \Delta x(f_1V_1^2 - f_2V_2^2)/D$$

A_c, A_x = cross-sectional flow area in the transverse and axial direction, respectively

D = hydraulic diameter of the rod bundle

f = friction factor

g = gravitational constant

K_s = empirical cross-flow correlation constant

ΔP = interbundle differential pressure

V_1, V_2 = axial velocity of water at the beginning of an increment

Δx = calculational increment in axial direction

ρ = fluid density.

Weisman[76] showed that the single-phase data indicating K to be a function of (u/V) could be explained by using a lateral momentum balance and recognizing that only a portion of the duct or bundle area is affected by cross flow. Weisman proposed that the pressure difference between channels i and j be obtained from

$$\left(\frac{A''}{a}\right)_2 a_2u_2 = \frac{g_c(a_2 - a_1)(\bar{P}_i - \bar{P}_j)}{\rho(\bar{V}_i + \bar{V}_j)} + \left(\frac{A''}{a}\right)_1 a_1u_1 , \qquad (3.75)$$

where

A'' = effective axial flow area over which the lateral velocity can be taken as u

a_k = lateral flow area from inlet to location k, where k = 1 or 2

P_i, P_j = pressures in adjacent channels at the edge of areas affected by cross flow.

Subscripts 1 and 2 indicate the inlet and outlet of the length step being considered.

The ratio (A''/a) was found to be a function of the length of the connecting region (slot length). The relationship found is shown in Fig. 3-10. For large slot lengths, A'' is considered the entire flow area, A, of the assembly. Equation (3.75) then simplifies to

$$(u_2 - u_1)A = \frac{g_c(a_2 - a_1)(\bar{P}_i - \bar{P}_j)}{\rho(\bar{V}_i + \bar{V}_j)} \quad . \tag{3.76}$$

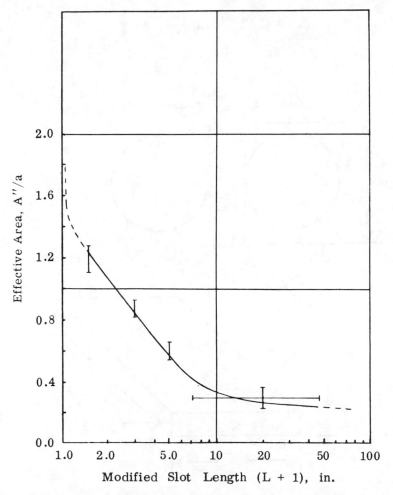

Fig. 3-10.   Effective area for assignment of lateral velocity in rod bundles. [From *Nucl. Technol.*, **15**, 467 (1972).]

Note that when $A'' = A$, $\bar{P}_i$ and $\bar{P}_j$ become the pressures at the outer edges of the two adjacent assemblies. Thus, if ΔP_l is the lateral pressure drop across one assembly, $\bar{P}_i - \bar{P}_j \simeq 2\Delta P_l$. After rearrangement, Eq. (3.76) can be rewritten as

$$(\Delta u/\Delta x)(\rho V_{av}) + g_c(\Delta P_l/\Delta y) = 0 \ , \tag{3.77}$$

where Δx is axial distance and Δy is lateral distance. This form is more convenient for estimating the lateral flow across a single assembly at an axial position removed from the core inlet.

The effective area concept is designed for use in determining flows between fuel assemblies. Computing cross flow between the subchannels of an assembly requires a different formulation of the lateral momentum balance. Rowe[77] writes the steady-state lateral momentum balance for the small gap between fuel rods (see Fig. 3-11) as

$$\frac{1}{g_c}\frac{\partial(Vw)}{\partial x} = \frac{s}{l}(P_i - P_j) - F \ , \tag{3.78}$$

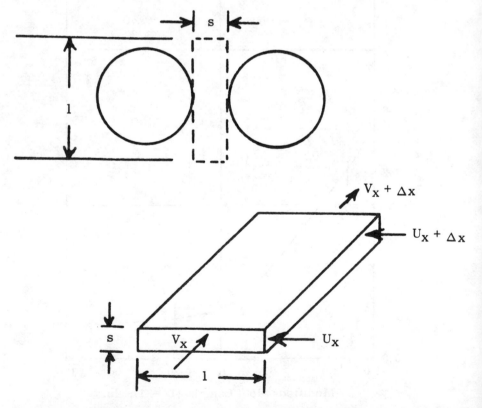

Fig. 3-11.   Control volume for lateral momentum balance. [From *Nucl. Technol.*, **15**, 466 (1972).]

where

x = axial distance in feet

F = frictional and form pressure loss

w = diversion cross flow between channels i and j, lb/s ft.

Other terms are defined in Fig. 3-11.

Rouhani[78] pointed out that Eqs. (3.75), (3.76), and (3.78) are simplifications of the general transverse momentum equation. Rouhani indicates that these simplifications are justified only when similar channels, in the absence of blockages, are considered. When blockages are present, Brown et al.[79] suggest that Eq. (3.78) can still be retained, providing a control volume of variable size is used. That is, length l is replaced by Δy, where Δy is a function of the distance above the blockage. With this modification, Brown et al.[79] were able to fit available data on lateral flow in the vicinity of a blockage placed in a simple two-channel geometry.

For flow blockage problems, Sha et al.[80] suggest the transverse momentum equation be expanded. This requires addition of the term $(1/g_c) [\partial(vw)/\partial x]$, where v is lateral velocity at right angles to w, to the left side of Eq. (3.78).

The lateral momentum balances of Eqs. (3.75), (3.76), and (3.78) implicitly assume that the two-phase mixture present in the core can be treated as a homogeneous fluid. The only direct experimental measurements on lateral two-phase flow are those of Madden[81] for air and water flow through slots. Madden defined separate discharge coefficients for gas and liquid streams. However, when these data were recalculated[82] in terms of a single discharge coefficient using an average density of the cross-flow stream, the data followed the curve, with some scatter, that Weisman[82] predicted for a 4-in.-long slot, giving some confirmation to homogeneous treatment.

Evidence[81,83] that the quality of the cross-flow stream may be somewhat higher than that of the main stream is available. However, quality variation is perhaps at most 0.1. Weisman[82] concluded that in view of the small size of the cross flow under most circumstances, such a variation generally would not lead to major error in the enthalpy of the main stream. Therefore, the homogeneous flow approximation almost universally used in subchannel calculations appears to be reasonable.

3-1.4 Sonic Velocity and Critical Flow

A sudden change in fluid velocity causes a large pressure change that is seen as a pressure wave, across which there is discontinuity in pressure and velocity. The pressure wave, which can be either a compression wave (high pressure behind a wave front) or a rarefaction wave (low pressure behind a wave front), propagates at sonic velocity relative to the flow. It can be set up by suddenly closing a valve in a flow line, by rupturing a high-pressure piping system (e.g., a LOCA), or by a steam burst in a reactor power excursion. The steam burst imparts kinetic energy, which is converted into a pressure pulse when the flow is interrupted. Pressure waves of this general nature are often called steam hammer or water hammer.

Pressure and velocity changes across an acoustic wave in one-dimensional flow can be derived from the momentum equation

$$\frac{\partial V}{\partial t} + V \frac{\partial V}{\partial x} = -\frac{1}{\rho} \frac{\partial P}{\partial x} , \qquad (3.79)$$

where V is fluid velocity, P is pressure, and x is distance. Since velocity change with respect to distance is relatively small in a straight pipe, the $\partial V/\partial x$ term is negligible, and Eq. (3.79) becomes

$$\frac{\partial V}{\partial t} = -\frac{1}{\rho} \frac{\partial P}{\partial x} . \qquad (3.80)$$

The pressure wave travels at the velocity of sound a_1; therefore, it will travel distance $(a_1 t)$ in time t. Let

$$\theta = x \pm a_1 t , \qquad (3.81)$$

where the negative sign is required for a wave moving in the direction opposite to V. By substituting Eq. (3.81) into Eq. (3.80), we obtain

$$\pm a_1 \frac{dV}{d\theta} = -\frac{1}{\rho} \frac{dP}{d\theta} . \qquad (3.82)$$

By integrating, this yields

$$\frac{P_2 - P_1}{V_2 - V_1} = \pm \rho a_1 . \qquad (3.83)$$

Equation (3.83) shows that in subcooled water with a density of 62 lb/ft^3 and a sonic velocity of 5000 ft/s, a 3 ft/s change in fluid velocity across the wave front causes a pressure pulse of 200 psi.

When a one-dimensional acoustic wave travels in a fluid of equal and opposite velocity, the wave becomes stationary with respect to the earth. We then have critical flow at this point (fluid at sonic velocity), and downstream pressure signals can no longer be transmitted to the upstream fluid. Thus, the critical flow rate is limited by the upstream and critical point conditions, but not by downstream conditions. In the case of a rupture of a PWR piping system, the critical flow rate at the break determines the severity of the accident.

In single-phase flow, sonic velocity a_1 and critical mass flow rate are directly and simply related:

$$a_1^2 = g_c v^2 / -(dv/dP)_s , \qquad (3.84)$$

where v is specific volume and subscript s indicates the derivative is evaluated at constant entropy. Critical mass flow rate G_c is given by

$$G_c^2 = a_1^2 \rho^2 = g_c / -(dv/dP)_s . \qquad (3.85)$$

Two-phase critical flow is, however, more complex: sonic velocity in a two-phase mixture varies with the amount of void present; velocity is usually measured at the wave front, where no phase change can take place instantaneously. On the other hand, critical flow rate is determined by the sonic velocity behind the wave front, where phase change does take place. This change depends on wave shape and bubble delay time which, in turn, depends on saturation pressure of the fluid. Therefore, it is necessary to consider sonic velocity and critical flows of two-phase mixtures separately.

3-1.4.1 Sonic Velocity

There are two limits for the response of a mixture of liquid and vapor to a pressure pulse: (a) mass transfer between the phases maintaining thermodynamic equilibrium, and (b) no mass transfer (frozen state) with liquid and vapor being independently isentropic. If no mass transfer is assumed, the computed velocity is often called the "frozen sonic velocity." If thermodynamic equilibrium is assumed, the computed velocity is substantially below frozen sonic velocity and the velocity actually observed at low and moderate pressures.

If vapor and liquid can be considered a homogeneous mixture, sonic velocity can be quite simply derived from Eq. (3.84) by using the mean density $\bar{\rho}$, where

$$\bar{\rho} = \alpha \rho_v + (1 - \alpha)\rho_l \ , \tag{3.86}$$

where α is void fraction and ρ_v, ρ_l are vapor and liquid densities, respectively. The frozen two-phase sonic velocity, a_{TP}, can be written as[84]:

$$a_{TP}^2 = \left[\frac{\alpha \bar{\rho}}{\rho_v a_v^2} + \frac{(1 - \alpha)\bar{\rho}}{\rho_l a_l^2} \right]^{-1} \ , \tag{3.87}$$

where a_v, a_l are sonic velocity in vapor and liquid, respectively. At low pressures where $a_l \gg a_v$, Eq. (3.87) becomes

$$\frac{a_{TP}^2}{a_v^2} = \left[\alpha^2 + \frac{\alpha(1 - \alpha)\rho_l}{\rho_v} \right]^{-1} \ , \tag{3.88}$$

Dvornichenko[85] derived a slightly different expression for the frozen sonic velocity

$$\frac{a_{TP}}{a_v} = \left[\sqrt{X} + \frac{(1 - X)\rho_v}{\sqrt{X}\rho_l} \right] \ , \tag{3.89}$$

where X is quality. Grolmes and Fauske[84] found that Eq. (3.87) fitted their low void fraction data very well. Therefore, it can be used to predict velocity of the sonic wave front in a low-void bubbly flow. At high void fractions, the assumption of a homogeneous mixture is no longer valid, and increasing deviations from Eq. (3.87) are observed.

In evaluating a_{TP} by either Eq. (3.88) or Eq. (3.89), it is not obvious what path should be used to evaluate a_v^2. If adiabatic behavior is assumed, then $a_v^2 = g_c P \gamma/\rho_v$, where γ is the ratio of specific heat at constant pressure to that at constant volume. If isothermal conditions are assumed, $a_v^2 = g_c P/\rho_v$.

Henry et al.[86] studied the situation at higher void fractions where slip must be considered. They used $\bar{\rho} = 1/[Xv_v + (1 - X)v_l]$ in Eq. (3.84) and expressed a^2 in terms of liquid density ρ_l, gas density ρ_v, and velocity slip ratio S. Equation (3.84) becomes

$$a_{TP}^2 = [(1 - X)\rho_v + X\rho_l]^2 \bigg/ \left[X\rho_l^2 \left(\frac{\partial \rho_v}{\partial P}\right)_{sat} + (1 - X)\rho_v^2 \left(\frac{\partial \rho_l}{\partial P}\right)_{sat} \right.$$

$$\left. - (\rho_l - \rho_v)\rho_l\rho_v \left(\frac{\partial X}{\partial P}\right)_s + X(1 - X)(\rho_l - \rho_v)\rho_l\rho_v \left(\frac{\partial S}{\partial P}\right)_s \right] , \qquad (3.90)$$

where subscript s indicates the derivative is evaluated at constant entropy. When it is assumed that no phase change occurs, Eq. (3.84) becomes

$$a_{TP}^2 = \frac{[(1 - X)\rho_v + X\rho_l]^2}{\left[X\rho_l^2 \left(\frac{\partial \rho_v}{\partial P}\right)_s + (1 - X)\rho_v^2 \left(\frac{\partial \rho_l}{\partial P}\right)_s + X(1 - X)(\rho_l - \rho_v)\rho_l\rho_v \left(\frac{\partial S}{\partial P}\right)_s \right]} . \qquad (3.91)$$

Equation (3.91) can be simplified for frozen annular flow (or stratified flow) as follows: For a stationary wave, Henry et al.[86] suggested

$$\frac{\partial S}{\partial P} = \frac{1}{a_{TP}} \left(\frac{\partial V_v}{\partial P} - \frac{\partial V_l}{\partial P}\right) . \qquad (3.92)$$

For the vapor phase,

$$\frac{\partial P}{\partial z} + \rho_v a_{TP} \frac{\partial V_v}{\partial z} = 0 .$$

And, for the liquid phase,

$$\frac{\partial P}{\partial z} + \rho_l a_{TP} \frac{\partial V_l}{\partial z} = 0 . \qquad (3.93)$$

Therefore,

$$\frac{\partial S}{\partial P} = \frac{-1}{a_{TP}^2} \left(\frac{1}{\rho_v} - \frac{1}{\rho_l}\right) .$$

By substituting Eq. (3.93) into Eq. (3.91),

$$a_{TP}^2 = \frac{\{[(1 - X)\rho_v + X\rho_l]^2 + X(1 - X)(\rho_l - \rho_v)^2\}}{\dfrac{X\rho_l^2}{a_v^2} + \dfrac{(1 - X)\rho_v^2}{a_l^2}} . \qquad (3.94)$$

For incompressible flow at low pressures ($a_l \gg a_v$), sonic velocity can be approximated as

$$\left(\frac{a_{TP}}{a_v}\right)^2 = 1 + \frac{1-\chi}{\chi}\left(\frac{\rho_v}{\rho_l}\right)^2 , \tag{3.95}$$

or, in case of no slip,

$$\left(\frac{a_{TP}}{a_v}\right)^2 = 1 + \frac{1-\alpha}{\alpha}\left(\frac{\rho_v}{\rho_l}\right) . \tag{3.96}$$

In high-quality or high void fraction flow, both Eqs. (3.95) and (3.96) indicate that $a_{TP} \rightarrow a_v$.

Henry et al.[86] compared their air-water mixture data with Eqs. (3.87) and (3.94). For the bubbly-flow region, they found sonic velocity agrees approximately with Eq. (3.87); for stratified flow (non-flowing system), Eq. (3.94) was applicable.

The stratified model of Eq. (3.93), which essentially gives the sonic velocity of vapor, is applicable for high-void steam-water (annular) flow. This appears to be verified by steam-water data obtained by Collingham and Firey,[87] England et al.,[88] Deich et al.,[89] and DeJong and Firey.[90]

Henry[91] subsequently observed that the velocity of a wave front at very low void fractions tends to approach that given by Eq. (3.87) with a_v^2 evaluated at isothermal conditions. For void fractions above those where Eq. (3.87) strictly applies, and below those of the annular region ($\alpha \leqslant 0.6$), Henry[91] correlated the available data by

$$a_{TP}/a_H = 1.035 - 1.671\alpha , \tag{3.97}$$

where

a_{TP} = actual propagation velocity in two-phase mixture

a_H = propagation velocity determined from Eq. (3.87) with a_v^2 evaluated under isothermal conditions.

The question of whether there is always negligible mass transfer at the wave front is not completely settled. On the basis of their data, Hamilton and Shrock[92] suggest that when vapor bubble size is very small, mass transfer may not be negligible. However, Henry[91] disagrees with their data interpretation. In the passage of a rarefaction wave, mass transfer may depend on the ability of new vapor bubbles to nucleate in the brief period of wave passage.

In the latter part of 1968, Edwards[93] reported his analytical study of a transient critical flow with flashing that was controlled by heat conduction. He suggested that the maximum time delay before a bubble forms decreases when system pressure increases as shown in Fig. 3-12. On the basis of Edwards' information, we expect that Grolmes and Fauske's[84] previously cited low-pressure sonic velocity data, which have an inherently long bubble delay time, would agree with the frozen model. Data at very high pressures where bubble delay time is short are unavailable.

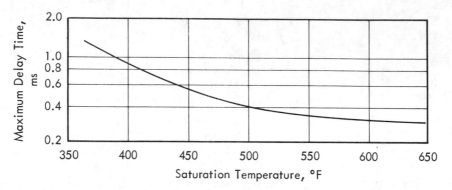

Fig. 3-12. Recommended bubble formation delay time (from Ref. 93).

The effect of phase change slows down the pressure wave behind the front and makes it flatter than obtained in a fully frozen system (e.g., air-water). Grolmes and Fauske[84] also observed this effect while studying the characteristics of both compression and rarefaction waves. Behind the wave front, the effects of heat and mass transfer become important and appear to prevent compression waves from steepening into shock waves, significantly lengthening the shape of the rarefaction waves. This effect has led Henry[91] to distinguish between the propagation velocity of a pressure pulse and the propagation velocity of a sound wave. Henry states that the frozen model holds strictly for a pressure pulse only. Propagation of a sound wave involves response to a continuous wave. At low frequencies, there is time for appreciable mass and heat transfer and, hence, lower propagation velocities can be expected. High frequency waves can be expected to have propagation velocities that approach those of a pressure pulse.

3-1.4.2 Critical Flow in Long Pipes

From the previous discussion, it is clear that there is an approach to thermodynamic equilibrium behind the wave front. Since critical flow is determined by conditions behind the front, some phase change must be considered. In a long pipe line, where there is adequate time for bubble nucleation and growth, thermodynamic equilibrium can be assumed.

The homogeneous equilibrium model is the simplest analytical model that can be postulated. The model assumes that

    1. both phases move at the same velocity

    2. the fluid is in thermal equilibrium

    3. flow is isentropic and steady.

By applying these assumptions to continuity and energy conservation equations, mass flow rate G is found by

$$G = [(2g_c J)(H_0 - H)/v^2]^{1/2} \ ,$$ (3.98)

where

H_0 = upstream reservoir enthalpy = $H + (G^2 v^2)/(2g_c J)$

H = local fluid enthalpy = $H_f - \chi H_{fg}$

v = local fluid specific volume = $(1 - \chi)v_l + \chi v_g$

J = thermal energy to mechanical energy conversion factor.

For fixed stagnation (upstream reservoir) conditions, the critical mass flow rate is obtained by finding the downstream pressure for which G exhibits a maximum. The results of these calculations for steam-water systems are presented in Fig. 3-13.

Moody[94] considered the available data for blowdown in long pipes. From an examination of nonequilibrium behavior during depressurization, Moody concludes that if pipe length is over five inches, equilibrium states can be expected. He found that available data on the blowdown of reservoirs through pipes of lengths >5 in. were predicted by Fig. 3-13 when calculations were based on reservoir conditions. This is in agreement with Sozzi and Sutherland's[95] conclusions, Caraher and DeYoung's[96] evaluation of semi-scale blowdown data, and Edwards'[93] observations.

Moody[94] points out that although homogeneous theory using reservoir conditions provides a good prediction of critical flow rates, it provides a poor prediction of pressure at the exit of the blowdown pipe. He concludes that mass fluxes are limited by homogeneous choking near the pipe entrance, but that a transition to slip flow occurs before reaching the exit. A second choked condition near the exit is produced and, if critical flow rates are to be evaluated on the basis of local conditions near the pipe exit, a slip flow model must be used. An alternative explanation may be that in the region adjacent to the exit, thermodynamic equilibrium may not yet be established.

Fauske[97] developed a "phases in equilibrium but separated" flow model for a long pipe which could be used with exit conditions. Fauske assumed:

1. Average velocities of different magnitude exist for each phase; i.e., slip flow is considered.

2. The vapor and liquid are in phase equilibrium throughout the flow path.

3. Critical flow is attained when the flow rate is no longer increased with decreasing downstream static pressure; i.e., $(\partial G/\partial P)_{H_0} = 0$.

4. The pressure gradient attains a maximum value for a given flow rate and quality.

In the absence of friction, the momentum equation for an isentropic annular flow can be written as

$$\frac{G^2}{g_c}\frac{dv}{dz} + \frac{dP}{dz} = 0 \ ,$$ (3.99)

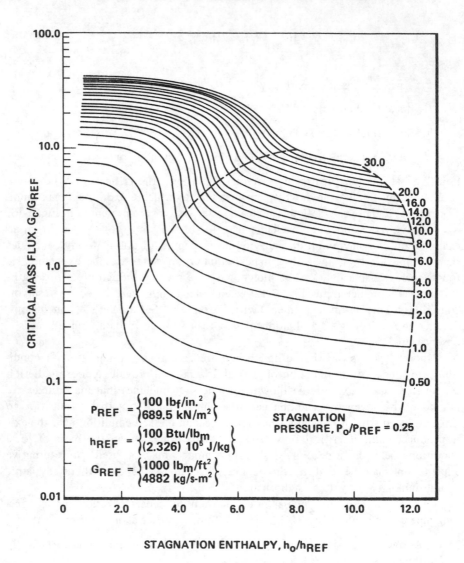

Fig. 3-13.   Critical mass flux—homogeneous equilibrium—steam-water mixtures. [Reproduced from NEDO-21052, General Electric Company (Sep. 1975).]

or

$$\frac{G^2}{g_c} = -\left(\frac{dP}{dv}\right)_s ,$$

(3.100)

where mean specific volume v is given as

$$v = \frac{\chi^2 v_v}{\alpha} + \frac{(1-\chi)^2 v_l}{1-\alpha} ,$$

(3.101)

and v_v and v_l are the specific volumes of vapor and liquid, respectively. By introducing slip ratio S, defined as

$$S = \frac{V_v}{V_l} = \frac{X}{1 - X} \frac{1 - \alpha}{\alpha} \frac{v_l}{v_v} \ , \tag{3.102}$$

Eq. (3.101) becomes

$$v = \frac{[(1 - X)v_l S + Xv_v][1 + X(S - 1)]}{S} \ . \tag{3.103}$$

Maximization of the pressure gradient is achieved from Assumption 4 above by varying the slip ratio; all other quantities are kept constant:

$$\frac{\partial v}{\partial S} = (X - X^2)\left(v_l - \frac{v_v}{S^2}\right) = 0 \ . \tag{3.104}$$

Hence, at critical flow, the slip ratio becomes

$$S = (v_v/v_l)^{1/2} = (\rho_l/\rho_v)^{1/2} \ . \tag{3.105}$$

The mass flow rate is obtained by substituting Eq. (3.103) into Eq. (3.100)

$$G^2 = \frac{-g_c}{\dfrac{d}{dP}\left\{[(1 - X)v_l S + Xv_v][1 + X(S - 1)]/S\right\}} \ , \tag{3.106}$$

or, by neglecting the insignificant term dv_1/dp,

$$G^2 = \frac{-g_c}{[(1 - X + SX)X]\dfrac{dv_v}{dP} + [v_v(1 + 2SX - 2X) + v_l(2XS - 2S - 2XS^2 + S^2)]\dfrac{dX}{dP}} \ . \tag{3.107}$$

Zivi[98] recognized that the work of accelerating the two phases to their ultimate velocities is a significant part of the total flow work. By assuming a thermal equilibrium in an annular flow and by neglecting wall friction, he obtained the maximum exit flow rate by maximizing the exit kinetic energy E of the flow

$$\frac{\partial E}{\partial \alpha} = \frac{G}{2}\left[2V_v \frac{\partial V_v}{\partial \alpha} X + 2V_l \frac{\partial V_l}{\partial \alpha}(1 - X)\right] = 0 \ . \tag{3.108}$$

Maximum kinetic energy is obtained when slip ratio S is given by

$$S = \frac{V_v}{V_l} = \left(\frac{\rho_l}{\rho_v}\right)^{1/3} \ . \tag{3.109}$$

Equation (3.109) agrees with that of Cruver and Moulton,[99] who used the criterion of maximum system entropy to determine critical flow conditions. On the other hand, Fauske[100] achieved reasonable agreement with low pressure experimental data using his critical flow equation, Eq. (3.107), and Armand's slip ratio,[101] where

$$\alpha = \frac{(0.833 + 0.167X)Xv_v}{(1 - X)v_l + Xv_v} \ . \tag{3.110}$$

Moody[94] argued that both Fauske's[97] work and his own earlier work,[102] which was based on energy conservation, were inconsistent. He derives a revised model that takes into account all the conservation laws. He finds that critical flow rate is given by the determinantal equation

$$\begin{vmatrix} b_1 & a_{12} & a_{13} \\ b_2 & a_{22} & a_{23} \\ b_3 & a_{32} & a_{33} \end{vmatrix} = 0 \ , \tag{3.111}$$

where

$$a_{11} = (G/g_c)v_m$$

$$a_{12} = (G^2/g_c)\partial v_m/\partial x$$

$$a_{13} = (G^2/g_c)\partial v_m/\partial S$$

$$a_{21} = (G/g_c)v_E^2$$

$$a_{22} = (G^2/2g_c)(\partial v_e^2/\partial x) + \partial H/\partial x$$

$$a_{23} = (G^2/2g_c)\partial v_e^2/\partial S$$

$$a_{33} = 0$$

$$a_{32} = \partial s/\partial x$$

$$b_1 = -(G^2/g_c)(\partial v_m/\partial P) + 1$$

$$b_2 = -[(G^2/2g_c)(\partial v_e^2/\partial P) + \partial H/\partial P]$$

$$b_3 = \left[\frac{\partial s}{\partial P} - \frac{1}{T} \left(\frac{m_{fg} + m_{gf}}{dP(GA)} \right) \left(\frac{u_v - u_l}{2g_c} \right)^2 dP \right]$$

$$u_v = v^*G$$

$$v^* = xv_v + (1 - x)Sv_1$$

$$A = A_v + A_l = \text{area}$$

$$x = \text{flowing quality} = u_vA_v/v_vGA$$

$$s = \text{entropy}$$

$$u_1 = \text{velocity of liquid}$$

$$v_v, v_l = \text{specific volumes of vapor and liquid, respectively}$$

$$v_m = v^* \left(x - \frac{1-x}{S} \right)$$

$$v_\epsilon^2 = (v^*)^2 \left(x - \frac{1-x}{S^2} \right)$$

T = absolute temperature

m_{fg}, m_{gf} = vaporization and condensation rates.

Equation (3.110) gives G as a function of P, x, and S. Moody concludes that critical G occurs where S is chosen so that

$$\begin{vmatrix} a_{11} & a_{12} & b_1 \\ a_{21} & a_{22} & b_2 \\ a_{31} & a_{32} & b_3 \end{vmatrix} = 0 \tag{3.112}$$

is satisfied. The results of Moody's computations are presented in Fig. 3-14 as the curves labeled "maximum slip ratio." Note that the critical slip ratio is not

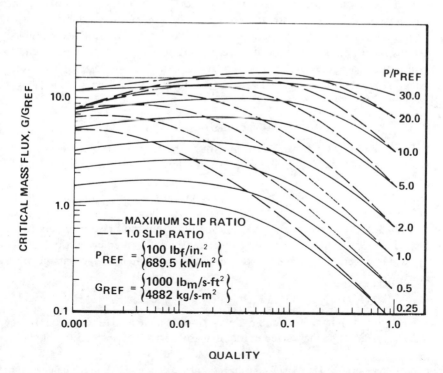

Fig. 3-14. Critical flow rate for water-steam using Moody's maximum slip ratio. [Reproduced from NEDO-21052, General Electric Company (Sep. 1975).]

constant for a given pressure, as it would be if it were a function of density ratio only, but it varies with quality.[a]

Moody[94] found good agreement between calculations of his model based on exit conditions and all data he examined, except those at very low qualities ($x < 0.02$) at $P = 100$ psia. There he found the measured data approached the mass fluxes calculated by his homogeneous model as quality was reduced.

In view of the uncertainty in the value of S to be used under break conditions, as well as possible nonequilibrium effects, it is suggested that computations of saturated blowdown rates through long pipes be based on reservoir conditions and the homogeneous model. For subcooled and low-quality blowdown at low pressures, Henry[86] suggests that an empirical correction factor, N, be used for correcting the homogeneous model as

$$G_c^2 = \frac{-g_c}{N\left(X\left.\frac{\partial v_v}{\partial P}\right|_s + v_v\left.\frac{\partial X}{\partial P}\right|_s\right)} , \qquad (3.113)$$

where $N = 20X$ for $P < 350$ psia, and $X \leqslant 0.02$.

3-1.4.3 Critical Flow in Short Pipes, Nozzles, and Orifices

When a reservoir discharges through a short pipe, orifice, or nozzle, the fluid can be subcooled just upstream of the break. Under these conditions, two types of choking mechanisms were seen by Zaloudek.[103] Zaloudek observed that an upstream choke can form at a vena contracta and a downstream choke can form as the back pressure is built up at the exit edge by flashing. He found that when choking occurred at the vena contracta, critical mass velocity was given by

$$G_c = C_1[2g_c\rho_l(P_{\text{upstream}} - P_{\text{sat}})]^{1/2} , \qquad (3.114)$$

where C_1 is an empirically determined constant found to range from ~0.61 to 0.64. When downstream choking occurred, the critical flow rate agreed with Burnell's[104] correlation.

Burnell[104] previously recognized the existence of a metastable state in the flow of flashing water through nozzles and hypothesized that water surface tension retarded the formation of vapor bubbles, thus causing the water to be superheated. He developed a semi-empirical method for predicting the flow of flashing water through square-edge orifices and correlated the discharge mass flow rate as

$$G_{\text{crit}} = \{2g_c\rho_l[P_{\text{upstream}} - (1 - C)P_{\text{sat}}]\}^{1/2} , \qquad (3.115)$$

where C is directly related to bubble delay time, which is a function of surface tension. The value of C that is recommended for design use is given in Fig. 3-15.

---

[a] The critical flows given in Fig. 3-14 for a slip ratio of 1.0 do not reduce to the homogeneous model critical flow given in Fig. 3-13. This appears to be indicative of an inconsistency in the analysis.

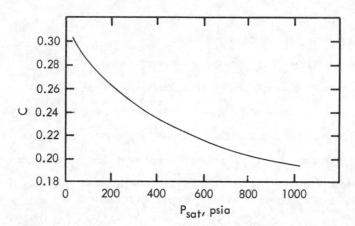

Fig. 3-15. Pressure coefficient for Burnell critical flow equation. [From *Engineering*, **164**, 572 (1947).]

The value of C determines the pressure undershot at the exit due to superheating the liquid.

Sozzi and Sutherland[95] also observed nonequilibrium effects in blowdown through short nozzles. They saw a strong dependence on nozzle length with the shorter nozzles giving higher flow rates. They found that fluid passing through a very short length will not have sufficient time to completely nucleate before leaving the tube. For nozzle lengths of about five inches, agreement between observations and homogeneous predictions were obtained. They also observed that critical mass flux decreased with increasing throat diameter.

Henry and Fauske[105] developed a model for critical flow in nozzles and short tubes which allows for nonequilibrium effects and considers a two-phase mixture upstream of the break. They proceed in a manner similar to that used by Fauske,[97] but do not assume thermodynamic equilibrium between phases. They use an empirical correlation to relate actual dx/dP to that (dx_e/dP) under equilibrium conditions. For a dispersed flow, they assumed

$$dx/dP = N(dx_e/dP) \ ,$$

(3.116)

where experimental parameter N is given by

$$N = x_e/0.14 \quad \text{for} \quad x_e \leqslant 0.14$$
$$N = 1.0 \quad \text{for} \quad x_e > 0.14 \ .$$

They concluded that the critical mass flow rate was given by

$$G_c^2 = \left[\frac{x_0 v_v}{nP} - (v_v - v_{lo}) \left\{ \frac{(1 - x_0)N}{s_{v\epsilon} - s_{l\epsilon}} \frac{dx_{l\epsilon}}{dP} - \frac{x_0 C_p [(1/n) - (1/\gamma')]}{P(s_{v0} - s_{lo})} \right\} \right]^{-1} \ ,$$

(3.117)

where

χ_0 = quality at stagnation conditions

n = polytropic exponent for vapor compression

γ' = isentropic exponent for vapor compression = C_p/C_v

C_p, C_v = specific heats of vapor at constant pressure and volume, respectively

$s_{v\epsilon}, s_{l\epsilon}$ = entropy under equilibrium conditions of vapor and liquid, respectively

s_{vo}, s_{lo} = entropy at stagnation conditions of vapor and liquid, respectively.

Equation (3.117) is coupled with the momentum equation describing the pressure history to obtain a solution in terms of stagnation conditions

$$(1 - \chi_0)v_{lo}(P_0 - P_t) + \frac{\chi_0\gamma'}{\gamma' - 1} (P_0 v_{vo} - P_t v_{vt}) = \frac{[(1 - \chi_0)v_l + \chi_0 v_{vt}]^2}{2} G_c^2 . \quad (3.118)$$

By substituting this equation into Eq. (3.117) and by rearranging leads to an equation, we get the ratio of throat pressure, P_t, to stagnation pressure, P_0.

$$\eta = P_t/P_0 = \left[\frac{\dfrac{1 - \alpha_0}{\alpha_0}(1 - \eta) + \dfrac{\gamma'}{\gamma' - 1}}{\dfrac{1}{2\beta\alpha_t^2} - \dfrac{\gamma'}{\gamma' - 1}} \right]^{\gamma'/(\gamma' - 1)} , \quad (3.119)$$

where

$$\beta = [(1/n) + (1 - v_{lo}/v_{vt})] \left[\frac{(1 - \chi_0)NP_t}{\chi_0(s_{v\epsilon} - s_{l\epsilon})_t} \frac{ds_{l\epsilon}}{dP} \right]_t - \frac{C_p[(1/n) - (1/\gamma')]}{(s_{vo} - s_{lo})}$$

$$\alpha_0 = \frac{\chi_0 v_{vo}}{(1 - \chi_0)v_{lo} + \chi_0 v_{vo}}$$

$$\alpha_t = \frac{\chi_0 v_{vo}}{(1 - X)v_{lo} + \chi_0 v_{vt}}$$

$$v_{vt} = v_{vo}(\eta^{-1/\gamma}),$$

and subscript t indicates throat conditions. Once η is obtained by solving the transcendental equation, G_c is obtained from Eq. (3.118).

Note that the correlation given for N holds only for a dispersed mixture of vapor and liquid. In the discharge of a saturated or subcooled liquid through an orifice or short tube, a separated flow pattern is often observed. The proposed correlation for N is not applicable under these conditions and Burnell's equation should be used for G_c prediction.

3-1.4.4 Decay of Pressure Waves

When a pressure wave reaches a rigid dead end, it is reflected as a wave of equal magnitude and sign; that is, a rarefaction wave is reflected as a rarefaction wave and a compression wave as a compression wave. When a wave reaches an open end or large reservoir, it is reflected as a wave of equal magnitude but of opposite sign. When a wave encounters a change in area, a portion of the wave is transmitted in the original direction of travel and a portion is reflected back. With the assumption of one-dimensional streamline flow, it can be shown[106] that the amplitude of the transmitted wave is given by

$$\frac{\Delta P_{transmitted}}{\Delta P_{incident}} = \frac{2A_{incident}}{A_{incident} + A_{transmission}} , \qquad (3.120)$$

and the ratio of reflected ΔP to the incident ΔP is

$$\frac{\Delta P_{reflected}}{\Delta P_{incident}} = \frac{A_{incident} - A_{reflection}}{A_{incident} + A_{reflection}} . \qquad (3.121)$$

This behavior is illustrated in Fig. 3-16.

In a frictionless system, the combination of forward and reflected waves is undamped and sets up a standing wave. In any real system, friction at the walls gradually reduces the amplitude of the waves. In a straight pipe, Lieberman and Brown[106] give decay with time t as

$$\Delta P = (\Delta P_0) \exp(-fVt/D) , \qquad (3.122)$$

where

D = diameter of the pipe

f = Fanning Friction Factor

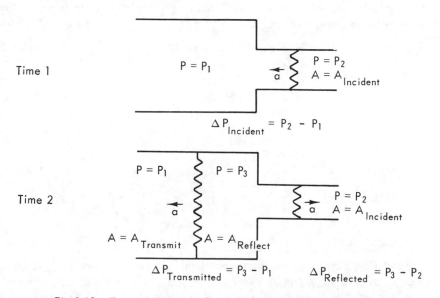

Fig. 3-16.   Transmission and reflection of acoustic wave at area change.

V = fluid velocity

$\Delta P_0 = \Delta P$ at zero time.

For a pipe diameter of 1 ft, a Fanning Friction Factor of 0.005, and water velocity of 20 ft/s, time constant D/fu is 10 s; thus, the frictional effect is relatively small.

When a pressure wave is reflected by a body that can significantly deflect, the amplitude of the reflected wave is appreciably reduced. The reader is referred to Streeter and Wylie[107] for discussion of this case. In a reactor system, the piping and major components deflect very little and can usually be considered as rigid; however, deflection of any thin wall internal structure should be considered.

The many area changes in a reactor loop lead to the appreciable attenuation of an acoustic wave as it proceeds through the system. However, the reflected waves set up at the area changes and the subsequent transmission and reflection of these reflected waves require an analysis that can follow this complex behavior in its entirety. The transmitted and reflected waves can act to reinforce each other in a particular region and to produce a high pressure across a particular component.

The arithmetic method[108] is conceptually the simplest procedure for establishing system behavior. The method neglects friction and generally assumes a horizontal system. The procedure basically requires that each wave be followed through the system. Pressure change across each wave is determined by using Eq. (3.83). When reflected and transmitted waves are created at area changes, their subsequent behavior must be followed. The pressure-time history at a given location is determined by summing the pressure changes that occur as the various waves pass through the location in question.

Tediously tracking individual waves makes the arithmetic method unsuitable for all but the simplest of systems. The method of characteristics[108] is most commonly used for solution of waterhammer problems. For single-phase flow, the continuity equation is written as

$$(\partial H/\partial t) + (a_1^2/g)(\partial V/\partial x) + V(\partial H)(\partial x) = 0 \; ; \tag{3.123}$$

the momentum equation is written as

$$g(\partial H/\partial x) + V(\partial V/\partial x) + (\partial V/\partial t) + (fV/2D)|V| = 0 \; , \tag{3.124}$$

where

H = fluid head

D = pipe diameter

x = axial distance

a_1 = sonic velocity

V = fluid velocity.

These partial differential equations are then converted into total differential equations that are subsequently expressed in finite difference form. The system is then divided into a series of control volumes and a set of finite difference equations is written for each control volume. Details of the numerical solution procedure are described in Sec. 4-1.3.

The simplest system modeling that can be used in this analysis is one that considers the reactor system to be one dimensional and, thus, to consist of a series of linear segments. While this is a reasonable way to simulate piping segments, it is not a good model of multidimensional regions within the reactor vessel. Streeter and Wylie[109] pointed out that a more realistic modeling of such regions can be obtained by using an equivalent interconnected piping network instead of a simple linear arrangement. Fabic[110] suggested further that when such network modeling is used, sonic velocities should be adjusted for the fact that piping network paths are longer than those actually traveled by the sonic waves. He proposed that the sonic velocities be multiplied by $\sqrt{2}$ and $\sqrt{3}$, respectively, for each two- and three-dimensional region. Comparing calculated and experimental results shows improved agreement when the equivalent piping network with modified sonic velocities is used.[111]

The pressures produced by wave propagation within a PWR primary coolant system are of greatest concern in the brief period while the system is still subcooled after a hypothetical LOCA. Once boiling begins, sonic velocity is greatly reduced and the pressure waves are damped out. The WHAM code[112] has been widely used for analyzing pressure wave behavior in subcooled reactor systems.

3-2 REACTOR VESSEL HYDRODYNAMICS

3-2.1 Flow in Plenums

Flow streams or vortices in the lower plenum can affect flow distribution at the core inlet. The downward jet formed by the flow coming down between the thermal shield and vessel wall causes a low-pressure region at the edge of the core inlet, which will reduce flow in the outer assemblies. Pressure at the center of a vortex with a horizontal plane of rotation is usually lower than the surrounding pressure; thus, the flow rate to any fuel assembly located above a vortex is reduced.

The most effective way to even off the inlet flow is to increase flow resistance at the bottom of the core. According to Prandtl's equation for velocity distribution improvement,[113]

$$\frac{\Delta V_{\text{after screen}}}{A_{\text{avg}}} = \frac{1}{K+1} \frac{\Delta V_{\text{before screen}}}{V_{\text{avg}}} , \qquad (3.125)$$

where

$$K = \frac{\Delta P_{\text{screen}}}{\rho V_{\text{avg}}^2 / 2g_c} . \qquad (3.126)$$

A similar relationship can be applied to the core inlet. For example, consider the situation where the Euler number at the core inlet is

$$\text{Eu} = \frac{\Delta P_{\text{core inlet}}}{\rho V_{\text{jet}}^2 / 2g_c} \leqslant 10 \ . \tag{3.127}$$

Under the most adverse conditions, K is then equal to 10. To obtain the pressure drop at the inlet of the hot channel, we add the stagnation pressure of the jet to the average inlet pressure drop. This leads to (ΔP of maximum flow channel/ ΔP of average channel) = 11/10. Since ΔP is proportional to V^2, the ratio of the maximum inlet flow rate to the average inlet flow rate is $(11/10)^{1/2}$.

Several analytical procedures based on potential flow theory have been used for predicting the velocity distribution in the lower plenum. Tong and Yeh[114] used conformal mapping to solve a two-dimensional approximation of the flow field. Yeh[115] obtained a solution for the potential field in the three-dimensional case. He divided the field into several regions (downcomer, lower plenum, and cylindrical region above bottom of downcomer) and solved the differential equations for each region by the separation of variables technique. Figure 3-17 shows the results of a typical calculation. This figure shows the benefit of a downcomer skirt that extends an appreciable distance below the core. While the flow is clearly not evenly distributed at the edge of the downcomer, distribution becomes more uniform as one moves upward.

Although potential flow theory can be considered useful in giving a qualitative view of behavior and an approximate estimate of velocities, analytical methods must simplify the geometry and ignore the many structural members in the lower plenum. Of course, any vorticity introduced cannot be considered. In view of the limitations of available analytical techniques, most reactor designers have determined velocity distribution in the lower plenum from experimental studies using scale models.

Hetsroni[116] describes hydraulically testing a model of a PWR vessel and core. Of the four dimensionless groups involved, geometry, relative roughness, Reynolds number, and Euler number, $[\Delta P(2g_c)/\rho V^2]$, he concludes that geometry and Euler number are most important. His tests used a $\frac{1}{7}$ scale model that preserved geometry. Care was taken to keep the Euler number similar in the prototype and model. The $\frac{1}{7}$ scale factor was also adjusted for relative roughness in small flow passages, but it was ignored in the large passages. The test results obtained a significantly improved flow distribution when a straight downcomer skirt extended below the core. A tapered skirt did not lead to satisfactory flow distribution. Flow distribution was also found to depend on the geometry of the lower plenum.

Application of the experimentally determined velocity distribution to an actual core design remains somewhat a problem. Under operating conditions where high power regions generate appreciable quantities of steam, the inlet flow redistributes to maintain a nearly constant pressure at the core outlet. Khan[117] suggested that the measured velocity distribution at the core inlet be used to determine an inlet pressure distribution. Further, he suggested that the inlet pressure distribution be assumed to hold under operating conditions. An alternative procedure would be to use inlet flow distribution to determine a set

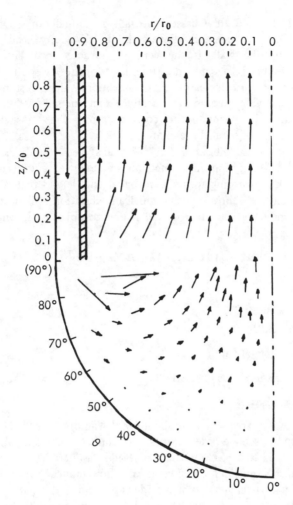

Fig. 3-17.   Typical velocity distribution in lower plenum—the arrows indicate direction and relative magnitude of fluid velocities. (From Ref. 115, courtesy of *Nucl. Eng. Des.*)

of flow resistances between the downcomer and various core assemblies. These flow resistances then might be assumed to be the same under operating conditions.

3-2.2 Flow-Induced Vibrations

3-2.2.1 *Hydraulic Vibration of Core Structure*

Vibration of a core structure can be caused by phenomena such as the periodic suction of a Venturi flow, by the wake of a vortex, by the impulse of an impinging jet, or by pulsating pressure.

When a high-speed flow passes through a gradually narrowing passage, the pressure head of the stream is converted into a velocity head, which creates a suction force on the wall. If the wall is movable or flexible, the passage is automatically reduced to zero and flow is stopped. As soon as flow becomes stagnant, steam pressure increases to its maximum value, pushing the wall back, thus providing a wide passage again. These suction and pushing forces can act periodically and, thus, vibrate the core structure. This mechanism appears to govern the vibration of parallel fuel plates.

Parallel fuel plate vibration has been analyzed by Miller,[108] Scavuzzo,[118] and Remick.[119] Miller applied a method of "neutral equilibrium" where the pressure and plate restoring forces were balanced. He was then able to determine a critical velocity, V, above which significant vibration will occur. He expresses this vibration-inducing velocity for fuel plate assemblies with long edges attached to side plates as

$$V_c = [15 \, g_c E (\Delta L)^3 d / \rho b^4 (1 - \gamma^2)]^{1/2} \ , \tag{3.128}$$

where

ΔL = plate thickness

b = plate width

d = flow channel depth

E = Young's modulus of elasticity

γ = Poisson's ratio.

Miller's results are frequently applied, and Wambsganss,[120] who extended the work to include second-order terms, concluded that critical flow velocities could be from 0.63 to 0.85 times those derived by Miller.

Miller and Kennison[121] analyzed the flow-induced vibration of a blade suspended in a flow channel, which widens at a point downstream of the leading edge—a condition similar to that observed by a control-rod blade. They concluded that there are three possible motions: (a) decaying vibrations, (b) divergent vibrations, and (c) static deflection against the channel wall. The motion observed is dependent on insertion length and flow rate. For short insertion lengths, there is a critical velocity above which divergent oscillations are obtained. They also conclude that a critical insertion length exists above which no oscillations can be sustained, but where deflection against the channel wall is possible. Wambsganss[122] concludes that the analysis explains experimental observations qualitatively, not quantitatively.

When a structural member receives the thrust of an impinging jet, it usually vibrates as the strength of jet varies. The worst vibration occurs when the natural frequency of the structural member equals that of the oscillating jet. Such jets can be caused by pumping pulses or vortex formation. According to Roshko[123] and Rouse,[124] a vortex at high Reynolds numbers (5×10^5 to 3.5×10^6) begins with laminar boundary layer separation and ends where the separated boundary layer becomes turbulent and the wake broadens further and becomes periodic. The vortices are shed at a frequency that is a function of the geometry of the body, spacing between adjacent bodies, and flow velocity. The shedding frequency is represented by the Strouhal number, St,

$$St = f_v L/V \ , \tag{3.129}$$

where

f_v = vortex shedding frequency

L = a characteristic dimension

V = stream velocity.

When the Reynolds number is in the range of 3.6×10^6 to 1×10^7, the vortex is in the transcritical range and the drag and Strouhal number remain constant at ∼1.2 as the velocity is changed.

If the source of the vortices or pressure pulses can be identified, the frequency of oscillations can be estimated. If the natural frequency of the structure is substantially above the frequency of the expected oscillation, vibration problems will not be significant. The response of complicated structures to an arbitrary dynamic excitation is difficult to predict. Bohm[125] developed an analytical model that approximates the internals as an interconnected beam-type structure and uses a transfer-matrix approach for determining natural frequencies. Bowers and Horvay[126] adopted a similar approach for estimating vibration frequencies of the core barrel and thermal shield. They noted that the narrow gap between the shell and vessel was responsible for a substantial increase in the virtual mass of the shell. They expressed their results in terms of ν, a nondimensional frequency parameter. The actual frequency of vibration, ω, is given by

$$\omega = (\nu/L)^2 (EI/m)^{1/2} \ , \tag{3.130}$$

where

L = length of the shell

$m = m_s + m_h$

m_s, m_h = mass per unit length of shell and entrained water, respectively

E = Young's modulus

I = moment of inertia of shell

M = attached mass (tubesheet, part of fuel, etc.).

They found ν to be a function of the weight ratio (mL/M), support stiffness, and bottom restraint. Typical values for ν were in the range of 1.5 to 2.2 for the fundamental mode; 3.4 to 4.2, 6.3 to 7.2, and 9.4 to 10.2 for the second, third, and fourth harmonics, respectively.

Actual operating experience indicates that the natural frequencies of the core barrel and thermal shields may not be significantly above the frequency of the exciting pulses. In some cores, damaging vibration of core barrel and shield have been encountered. More sophisticated approaches to this problem are required.

Most current (1976) analyses of core barrel and thermal shield vibration problems are based on the assumption that the exciting force is due to pump-induced pulsating coolant pressure. Experimental measurements of the magnitude and frequency of this forcing function are needed for predicting structural response.

Penzes[127] shows how a known pulsating pressure at the inlet to downcomer region can be used to determine the pulsating pressure distribution on the cylindrical core barrel or thermal shield. Penzes introduces the concept of a time-dependent body force (time-dependent additional thickness) in the governing differential equation. With this conceptual substitution for actual loading, the time-dependent mixed boundary value problem can be represented by a vibration problem with homogeneous boundary conditions. Bowers and Horvay[128] applied Penzes' approach to predicting internals response, but obtained vibration amplitudes considerably smaller than those predicted by Penzes.

The most sophisticated approach devised is that of Au-Yang.[129] He observes that flow-induced vibration is a stochastic process. Since stress and strain within a structure can be obtained from displacement distribution, Au-Yang develops expressions for predicting rms displacement and displacement power spectral density as a function of location. The prediction requires a knowledge of the natural frequency and mode shapes of the structure, virtual mass, pressuring loading function, and damping ratios. The loading function and damping ratios must be developed on the basis of experimental information (Penzes' analysis could be used to determine loading function if pressure fluctuations are known). In-fluid frequencies and modal shapes can be predicted by currently available finite element computer programs providing the virtual mass is supplied to the program. The virtual masses, in-fluid natural frequencies, mode shapes, and experimentally-based loading functions and damping ratios are used to compute displacement power spectral density in terms of Powell's acceptances.[130]

The complexity of most reactor coolant systems makes it difficult to predict the frequency and magnitude of hydraulic forcing functions; therefore, scale models of system components are often tested under simulated operating conditions.

3-2.2.2 Hydraulic Vibration of Fuel Rods

The hydraulic vibration of cylindrical fuel rods,[b] with their axes parallel to the flow, has been studied by Burgreen et al.,[131] Quinn,[132] Sogreah,[133] Pavlica

---

[b]Hydraulic vibration of the tubes in steam generators is also of concern to the designers. One possible cause of such vibration is the flow of liquid from the entrance ports across the lower portion of the tube bundle. Gorman [*Nucl. Technol.*, **61**, 324 (1976)] indicates that cross-flow-induced vibration from this source may be prevented by keeping the Strouhal number below a critical value which is a function of tube damping, tube mass per unit length, tube diameter, and liquid density.

and Marshall,[134] Paidoussis,[135] and Morris.[136] The earliest correlation is that due to Burgreen, who found that single rods vibrate at their natural frequency (reduced slightly by viscous damping in water) and that amplitudes of the vibrations are given by

$$\frac{\delta}{D} = (0.83 \times 10^{-10})^3 C^3 \left(\frac{\rho V^2}{\mu \omega}\right)^{0.77} \left(\frac{\rho_1 V^2 L^4}{EI}\right)^{0.42}, \qquad (3.131)$$

where

δ = maximum rod deflection

D = rod diameter

C = 5 for pin-ended rod, 1 for rigidly held rod, and 2.08 for one end rigidly held, the other pin ended

ω = the natural frequency of vibration

ρ_1 = rod density

E = Young's modulus of elasticity

I = moment of inertia

V = axial velocity

L = rod length.

Paidoussis[135,137] extended the work of Burgreen et al.[131] His most recent correlation for deflection amplitudes is

$$\frac{\delta}{D} = \alpha_1^{-4} \left[\frac{u^{1.8}\left(\frac{L}{\alpha}\right)^{1.8} \mathrm{Re}^{0.25}}{1 + u^2}\right] \left(\frac{D_e}{D}\right)^{0.4} \left(\frac{B^{2/3}}{1 + 4B}\right) (5 \times 10^{-4} \times K), \qquad (3.132)$$

where

α_1 = first-mode beam eigenvalue

$B = M + m/M$, M = rod mass, m = fluid mass

$u = (M/EI)^{1/2}(VL)$

V = axial velocity

D_e = hydraulic diameter of the flow channel

K = constant that is 1 for quiet flow and 5 for realistic disturbance levels.

This correlation shows much less dependence on fluid viscosity than Paidoussis' original proposal. It brings it into close agreement with the data of Quinn[132] and Pavlica and Marshall,[134] which show almost no fluid temperature effect.

Paidoussis[137] also examined the vibration of 19-element wire-wrapped bundles and found vibration amplitudes differed substantially from those of Eq. (3.131) and that they did not increase significantly with flow. Kinsel[138] studied the vibration characteristics of 217-pin spiral wire-wrapped fuel assemblies in water. He observed very low water pin displacements, generally 5 mm or less. The 0.23-in.-diam pins showed distinct resonances at low flows, but broad-band Gaussian distributions at high flows.

Pavlica and Marshall[134] examined the vibration of rod bundles with spring clip spacers. In correlating the data, they used the stiffness (EI) of the bundle rather than that of the individual rods. Since intermediate rod-to-rod connectors are not rigid, they used experimental values of (EI) rather than an estimate of single-element parameters. With this change, they found their room temperature data agreed with the original Paidoussis correlation. The discrepancy between their 150°F data and correlation appears to be remedied by the reduced viscosity effect in the revised correlation.

Gorman[139,140] investigated the effect of void fraction on single-pin vibration in both an annulus and a seven-rod bundle. Pin dimensions and bundle arrangement were similar to those found in CANDU designs. Gorman found that rms pin displacement increased substantially under two-phase conditions. The maximum rms displacement occurred at simulated qualities in the range of 12 to 20%. At a constant mass flow, the maximum rms displacement under two-phase conditions in a rod bundle could be up to eight times that observed in single-phase flow.

3-2.3 Hydraulic Design of Control Rods

To obtain a simplified method of analysis for estimating the drop time of a control rod, the basic assumption is made that all the hydraulic forces are proportional to the square of the rod velocity. The equation-of-motion of the control rod is then

$$\frac{M+m}{g_c} \frac{dV}{dt} = \Sigma F \ , \qquad (3.133)$$

where M is the mass of control rod and m is the virtual mass of the body due to fluid acceleration; i.e.,

$$m = M\rho_l/\rho_M \ . \qquad (3.134)$$

The equation-of-motion can be simplified to

$$\frac{dV}{dt} = C_2 - C_3 V^2 \ . \qquad (3.135)$$

Solution of the above equation is

$$t = \frac{1}{2(C_2 C_3)^{1/2}} \ln \frac{1 + [1 - \exp(-2C_3 x)]^{1/2}}{1 - [1 - \exp(-2C_3 x)]^{1/2}} \ , \qquad (3.136)$$

where x represents distance traveled.

Asymptotically, as $x \to \infty$ (or $x \gtreqqless 2/C_3$),

$$t = \frac{C_3 X + \ln 2}{(C_2 C_3)^{1/2}} . \tag{3.137}$$

At the end of the rod's travel, it enters a dashpot with a small clearance between it and the rod. This clearance produces a high frictional force that rapidly decelerates the rod. The resisting force in the dashpot is linearly proportional to the distance of insertion, Z. The equation-of-motion when the rod enters the dashpot is

$$\frac{dV}{dt} = C_2 - (C_3 + C_4 Z)V^2 . \tag{3.138}$$

The solution to this equation is

$$V^2 = V_0^2 \exp(-2C_3 Z - C_4 Z^2) + 2 \exp(-2C_3 Z - C_4 Z^2)$$

$$\times C_2 \int_0^Z \exp(2C_3 Z + C_4 Z^2) dz , \tag{3.139}$$

where V_0 is the entering velocity. This can be rearranged to

$$V^2 = V_0^2 \exp(C_3^2/C_4) \exp(-u^2)$$

$$+ (2C_2/\sqrt{C_4}) \exp(-u^2) \int_{C_3/\sqrt{C_4}}^{u} \exp(u^2) du , \tag{3.140}$$

where

$$u = C_3/\sqrt{C_4} + \sqrt{C_4} Z .$$

Hence, rod velocity in the dashpot can be evaluated at each Z.

3-3 FLOW INSTABILITY

The term "flow instability" refers to flow oscillations of constant or variable amplitude that are analogous to vibrations in a mechanical system. Mass flow rate, pressure drop, and voids can be considered equivalent to the mass, exciting force, and spring of the mechanical system. In this connection, the relationship between flow rate and pressure drop plays an important role. Flow oscillations can be aggravated when there is thermohydrodynamic coupling between heat transfer, void, flow pattern, and flow rate; however, oscillations can occur even when the heat source is held constant. Both hydrodynamic and thermohydrodynamic instabilities are discussed, but the more elaborate nuclear thermohydrodynamic instabilities of boiling channels in water-cooled reactors are beyond the scope of this book.

Flow oscillations are undesirable in boiling, condensing, and other two-phase flow devices for several reasons:

1. Sustained flow oscillations can cause undesirable forced mechanical vibration of components.

2. Flow oscillations can cause system control problems of particular importance in liquid-cooled nuclear reactors where the coolant (such as water) also acts as a moderator.

3. Flow oscillations affect local heat transfer characteristics and boiling crisis.

Thus, the critical heat flux (CHF) was found by Ruddick[141] to be reduced 40% when the flow was oscillating. This adverse effect was also found by Lowdermilk.[142] Mayinger et al.[143] found a similar reduction of CHF in an oscillating water flow at 2000 psia.

3-3.1 Nature of Various Flow Instabilities

3-3.1.1 Flow Pattern Instability

This occurs when flow conditions are close to the point of transition between bubbly flow and annular flow. A temporary increase in bubble population in bubbly-slug flow (arising from a temporary reduction in flow rate) can change the flow pattern to annular flow with its characteristically lower pressure drop. When the driving pressure drop over the channel remains constant, flow rate attains a greater value. However, as flow rate increases, the vapor generated (even for unchanged heat transfer characteristics) can become insufficient to maintain annular flow. The flow pattern then reverts to that of bubbly-slug flow and the cycle is repeated. The low-pressure drop characteristics of annular flow have been experimentally demonstrated by Wallis.[144] The cyclic behavior is partly due to the delay incurred in acceleration and deceleration of the flow.

3-3.1.2 Ledinegg Instability

Consider the operation of a boiler tube to which there is a constant heat input. At low flow, the boiler exit quality and velocity are high since the fluid is entirely, or almost entirely, vapor. This high vapor velocity accounts for much of the pressure drop. When the flow is increased, slightly more vapor is generated and the pressure drop increases. However, as flow increases further, exit quality begins to decrease and velocity at the tube exit decreases. The pressure drop decreases and continues to decrease until the tube contents are all water. The pressure drop then increases as the water velocity increases. Experimental observations of this phenomenon are shown in Fig. 3-18 (Ref. 145).

The effect of this phenomenon on stability can be understood by reference to Fig. 3-19, where we consider the operation of a boiling channel with a constant exit pressure. A portion of the pressure versus flow curve, which occurs at very low flow rates, is shown at the left. The corresponding curve for the boiler feed system is also shown. The system will operate where the two curves

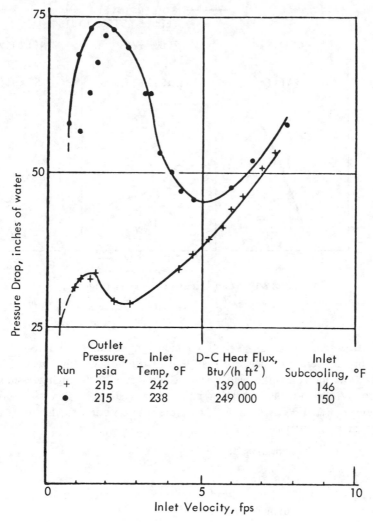

Fig. 3-18.   Static pressure drop through a 0.174-in.-i.d. stainless-steel AISI Type-347 heated single tube (from Ref. 145).

intersect. If flow to the boiler increases by a small amount, ΔW, boiler pressure becomes greater than supply pressure. This decelerates the flow and the system returns to the original operating point. On the other hand, if $\partial(\Delta P)/\partial G$ is negative, as shown in the central portion of Fig. 3-18, then an increase in flow decreases the inlet pressure so that the supply flow increases and continues to increase. This flow excursion can be predicted by the Ledinegg criterion,[147] which states that a necessary condition for the excursion is $\partial(\Delta P)/\partial G < 0$, where the steady-state characteristic of channel $G(\Delta P)$ is not single valued with identical inlet enthalpy and heat addition.

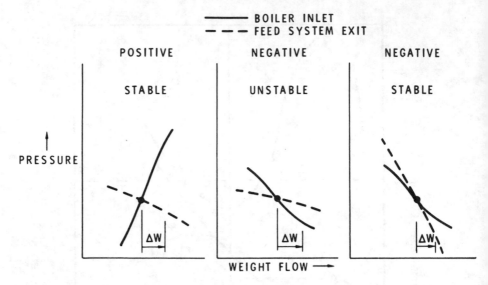

Fig. 3-19.   Boiler resistance (from Ref. 146).

The flow can be stabilized in this region by providing a pump-supply curve whose slope is even more negative than that of the boiler (right side of Fig. 3-19). If boiler flow increases, the feed system cannot supply the high pressure needed and flow returns to the operating point. The system is then statically stable even though $\partial(\Delta P)/\partial G$ is negative. Therefore, the criterion for Ledinegg or static stability can be written as

$$\frac{\partial(\Delta P)_{\text{feed}}}{\partial G} - \frac{\partial(\Delta P)_{\text{loop}}}{\partial G} < 0 \ , \tag{3.141}$$

where $(\Delta P)_{\text{feed}}$ is the driving head supplied by the feed pump and $(\Delta P)_{\text{loop}}$ is the flow friction head demanded by the loop.

Stenning and Veziroglu[148] call this type of flow oscillation "pressure drop oscillation." Generally, it can be avoided by increasing the pressure drop at the channel inlet or by using a pump with a large head reduction at high flow rates.

3-3.1.3 Density Wave or Dynamic Instability

A temporary reduction in the inlet flow to a boiling channel increases the rate of evaporation, thereby raising the average void fraction. This disturbance affects elevation, acceleration, and frictional pressure drop, as well as heat transfer behavior. For certain conditions of channel geometry, thermal properties of the heated wall, flow rate, inlet enthalpy, and heat flux, resonance with sustained oscillation can occur. Stenning and Veziroglu[148] found that the frequency of a density wave oscillation is higher than that of a pressure drop

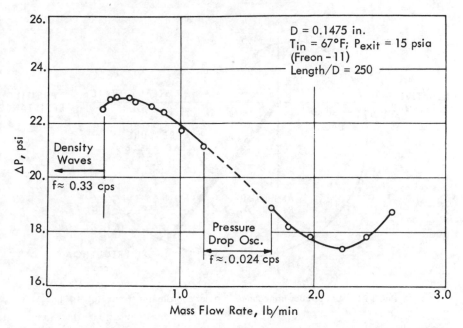

Fig. 3-20. Density and pressure wave oscillation regions.[147] [From *Proc. 1965 Heat Transfer and Fluid Mechanics Institute*, Stanford University Press, Stanford, California (1965).]

oscillation (Fig. 3-20). Neal and Zivi[149] recognized that this oscillation is caused by flow-void feedback when a 180-deg phase shift exists between a flow disturbance and its void volume response. This can occur even when the boiler is operating in a region where $\partial(\Delta P)/\partial G$ is positive. We consider a single, closed loop containing pump, boiler tube with constant heat input, and condenser and assume that some sinusoidal flow oscillation at a specified frequency is introduced. If pressure drop across the boiler is examined, pressure oscillations are observed which are not necessarily in-phase with the flow oscillation; they can lag the flow oscillations.

We assume that this process is repeated for a number of different imposed frequencies. At each frequency, the lag (phase angle) between the velocity and pressure oscillations is determined as well as the ratio of amplitude of flow oscillation A_F to the amplitude of the pressure oscillation, A_P. A polar coordinate plot, usually called a Bode diagram, of a typical set of results is shown in Fig. 3-21 (Ref. 146). At steady state (zero frequency), pressure and flow are in phase, and the boiler has a completely positive resistance. Where the curve crosses the horizontal axis again, pressure and velocity are out-of-phase by 180 deg, and the boiler has a completely negative resistance.

To determine whether instability will occur in our system, we must perform a similar experiment with our feed system and expect to get results such as those

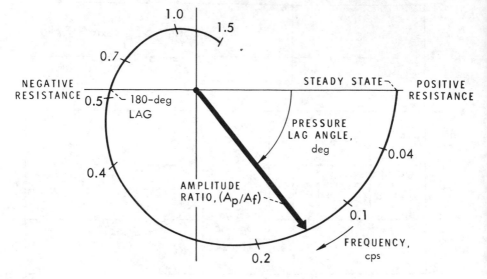

Fig. 3-21.   Boiler-inlet impedance in polar coordinates (from Ref. 146).

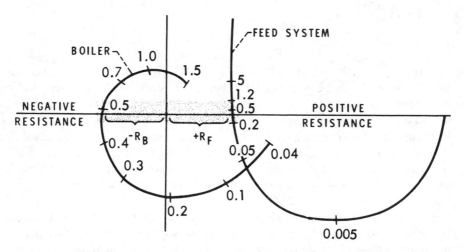

Fig. 3-22.   Boiler-inlet and feed-system impedances (from Ref. 146).

in Fig. 3-22 (Ref. 146). The feed system has a positive resistance at all frequencies. For the frequency range where the boiler has a negative resistance, $-R_B$, the feed system has a positive resistance, R_F. As long as $R_F > R_B$, the system is stable. If this is not the case, dynamic instability results.

3-3.1.4 Parallel Channel Instability

Parallel channel instability can arise when a number of channels are connected at common headers. Under these conditions, total flow can remain constant, but oscillations in the flow to some channels can occur. This problem is of particular concern in a pressure tube reactor where parallel channels are connected only at the inlet and exit. The problem can also arise in once-through steam generators. Moxon[150] notes that parallel channel instability is seen in such generators when a pressure drop perturbation arising from the evaporator and superheater is roughly equal to and 180-deg out-of-phase with a perturbation originating at the inlet.

3-3.1.5 Thermal Oscillation

This oscillation occurs in a film boiling region during rapid heating of a cryogenic system. It is a thermal response of the vapor film to flow disturbances. Disturbances that can be reinforced by a mechanism such as acoustic resonance become dominant in the system; these oscillations are often called acoustic instabilities. The velocity of sound is greatly reduced by the presence of small quantities of void; this can lead to "pipe organ" resonances in short pipe lengths.

Observed frequencies have been related to the harmonics of a pipe that is open at both ends, and to Helmholtz resonance. This phenomenon has been reported by Edeskutz and Thursten.[151] The inception of oscillations can be predicted by

$$q''/GH_{fg} = 0.005 \, (v_l/v_v) \; , \qquad\qquad (3.142)$$

where H_{fg} is enthalpy change on vaporization and the range of (v_v/v_l) is 9 to 25.

Cornelius and Parker[152] detected acoustic instability (5 to 30 cps) of refrigerant Freon-114 in a natural circulation loop at a bulk temperature below critical temperature, but at critical pressure. Walker and Harden[153] suggested that the region of pressure and flow fluctuations associated with a natural circulation loop operating at supercritical pressures occurs near maximum in a plot of density enthalpy product versus fluid temperature. Bergles et al.[154] examined thermal oscillation in boiling water flow at 1000 psia. They found oscillations characterized by a frequency >35 cps and most pronounced in the subcooled region. They also found pressure-drop amplitudes very large compared to steady-state values. The oscillations are considered to be acoustic in nature and induced by the collapse of subcooled voids.

3-3.2 Effects of Various Parameters on Flow Instability

The effects of various flow-instability parameters, observed in studies previously mentioned, are

1. High heat input aggravates flow instability because of the resulting high rate of vaporization.

2. Increased inlet subcooling within a critical value reduces stability. Beyond that value, a further increase in subcooling can reverse the trend and stabilize flow.

3. Increased resistance at the exit strongly decreases stability since it increases exit pressure drop when a large void is generated in the channel.

4. Increased resistance at the inlet strongly increases stability. When flow is slowed by a disturbance, a large pressure head is retained at the inlet and is available for forcing the fluid through the channel, thereby stabilizing the flow.

5. Increased system pressure increases stability in proportion to the increase in reduced pressure, primarily because the difference in specific volumes of vapor and liquid then becomes smaller.

6. Increased static head in a long vertical channel increases the stability of upward boiling flow and decreases the stability of downward boiling flow. This occurs because the buoyancy of the vapor aids upward flow and opposes downward flow. With the latter, there is a tendency for the vapor to agglomerate and remain in the channel.

3-3.3 Theoretical Analysis

Acoustic or thermal oscillations are primarily associated with startup and transient situations[149] and are usually not the major concern of the designer. Flow-pattern instabilities are imperfectly understood. Experimental observations are recommended. At present, the designer directs his major consideration to Ledinegg and dynamic instability. Analysis for both problems can be carried on simultaneously since Ledinegg instability can be considered a special dynamic case occurring at zero frequency. These problems are fairly well understood and a theoretical analysis can be made.

3-3.3.1 Basic Thermohydrodynamic Equations

Tong[23] reviewed this area extensively; we follow his discussion. First, we write the conservation equations for one-dimensional flow, with slip, in a vertical channel. For the continuity balance, we have

$$\frac{\partial}{\partial t} [\rho_l(1 - \alpha) + \rho_v \alpha] + \frac{\partial}{\partial z} [\rho_l(1 - \alpha)V_l + \rho_v \alpha V_v] = 0 \ , \qquad (3.143)$$

where z indicates axial distance and V_l and V_v are the liquid and vapor velocities, respectively. For the energy balance, we have

$$\frac{\partial}{\partial t} [\rho_l(1 - \alpha)H_l + \rho_v \alpha H_v] + \frac{\partial}{\partial z} [\rho_l(1 - \alpha)V_l H_l + \rho_v \alpha V_v H_v] = q''' \ , \qquad (3.144)$$

where energy dissipation is neglected, q''' is the heat input per unit volume of fluid and H_l and H_v are the liquid and vapor enthalpies, respectively. The momentum balance can be written as

$$\frac{1}{g_c}\frac{\partial}{\partial t}[\rho_l(1-\alpha)V_l + \rho_v\alpha V_v] + \frac{1}{g_c}\frac{\partial}{\partial z}[\rho_l(1-\alpha)V_l^2 + \rho_v\alpha V_v^2] + \frac{\partial P}{\partial z} + F$$

$$\pm [\rho_l(1-\alpha) + \rho_v\alpha]\frac{g}{g_c} = 0 \ . \tag{3.145}$$

A positive sign for the final term indicates upward flow; a negative sign indicates downward flow; F is the frictional force per unit volume of fluid.

These conservation equations assume that mixture behavior is adequately defined by overall conservation equations obtained from summing the conservation equations for the individual phases. Furthermore, each individual phase is assumed to occupy a given fraction of the channel cross section and to have a velocity that is constant across the cross section occupied.[c]

If the system remains at a nearly constant pressure, the equation-of-state can be written as

$$\rho_l \approx \rho_l(H_v) \tag{3.146}$$

$$\rho_v \approx \rho_v(H_v) \ . \tag{3.147}$$

Note that in subcooled boiling, the mixture density should be calculated on the basis of the amount of local boiling void present (see Sec. 3-1.2).

The preceding equations can be simplified by introducing the quality defined as

$$X = \frac{w_v}{w_v + w_l} = \frac{\alpha\rho_v V_v}{\alpha\rho_v V_v + (1-\alpha)\rho_l V_l} \ . \tag{3.148}$$

If local mass velocity is defined as

$$G = \alpha\rho_v V_v + (1-\alpha)\rho_l V_l \ , \tag{3.149}$$

the continuity balance becomes

$$\frac{\partial\bar{\rho}}{\partial t} + \frac{\partial G}{\partial z} = 0 \ . \tag{3.150}$$

The volume-weighted mean density, $\bar{\rho}$, is given by Eq. (3.20) and the energy balance is given by

$$\bar{\rho}\frac{\partial H}{\partial t} - H_{fg}\frac{\partial\psi}{\partial t} + G\frac{\partial H}{\partial z} = q''' \ . \tag{3.151}$$

Quantity ψ, defined by

$$\psi = \rho_l X(1-\alpha) - \rho_v\alpha(1-X) \ , \tag{3.152}$$

is a slip correction for the energy balance and is zero for nonslip flow. Mixing cup enthalpy H is defined by

$$H = H_l + X(H_v - H_l) \ . \tag{3.153}$$

---

[c]Two-phase flow conservation equations have been written in a variety of forms. For a more comprehensive view of this subject, the reader is referred to Refs. 155, 156, and 157.

The momentum balance is

$$\frac{1}{g_c}\left[\frac{\partial G}{\partial t} + \frac{\partial}{\partial z}(\nu' G^2)\right] + \frac{\partial p}{\partial z} + F \pm \frac{\bar{\rho}g}{g_c} = 0 \ , \tag{3.154}$$

where ν', the "effective specific volume for spatial acceleration," is

$$\nu' = \frac{1}{G^2}[\rho_l V_l^2(1-\alpha) + \rho_v V_v^2 \alpha] = \frac{\chi^2}{\alpha \rho_v} + \frac{(1-\chi)^2}{(1-\alpha)\rho_l} \ . \tag{3.155}$$

This becomes $\nu' = 1/\bar{\rho}$ for nonslip flow, and the equation-of-state can be written as

$$\bar{\rho} = \bar{\rho}(H) \ . \tag{3.156}$$

If channel size is small (e.g., between fuel rods in a PWR core) and vapor quality is sufficiently low so that bubbly flow exists, slip is negligible at high pressures. This is particularly true during a flow oscillation, since void distribution changes as flow changes from accelerating to decelerating.

3-3.3.2 Linearization and Laplace Transformation

When we consider a steam generator consisting of a large number of boiler tubes, each generating approximately the same amount of steam, it is usually adequate to consider the steam generator as a single channel. If we utilize the technique of small perturbations, the variables in Eqs. (3.143), (3.144), and (3.145) can be described by

$$G(z,t) = G(z,0) + \Delta G(z,t)$$

$$H(z,t) = H(z,0) + \Delta H(z,t)$$

$$\bar{\rho}(z,t) = \bar{\rho}(z,0) + \Delta\bar{\rho}(z,t)$$

$$q'''(z,t) = q'''(z,0) + \Delta q'''(z,t) \ , \tag{3.157}$$

where $\Delta G(z,t)/G(z,0) \ll 1$, etc. If these variables are substituted into the conservation equations and if the second-order terms are ignored, the equations can be linearized to yield

$$\frac{\partial \Delta G}{\partial z} = -\frac{\partial \Delta\rho}{\partial t} \tag{3.158}$$

$$\Delta q''' = G\frac{\partial \Delta H}{\partial z} + \Delta G\frac{\partial H}{\Delta z} + \rho\frac{\partial \Delta H}{\partial t} \tag{3.159}$$

$$\frac{1}{g_c}\left[\frac{\partial \Delta G}{\partial t} + \frac{\partial}{\Delta z}\left(\frac{2G\Delta G}{\rho} - \frac{G^2}{\rho^2}\Delta\rho\right)\right]$$

$$+ \frac{\partial \Delta P_{\text{channel}}}{dz} + \Delta\rho\frac{g}{g_c} + \frac{fG}{g_c D_e \rho}\Delta G - \frac{fG^2}{2g_c D_e \rho^2}\Delta\rho = 0 \tag{3.160}$$

$$\Delta\rho = \frac{\partial\rho}{\partial H}\Delta H + \frac{\partial\rho}{\partial G}\Delta G + \frac{\partial\rho}{\partial q'''}\Delta q''' \ . \tag{3.161}$$

A Laplace transformation of the above is used to obtain a transfer function of the inlet flow variation with respect to a variation of heat input as

$$G(s) = \frac{\Delta G(s)}{\Delta q'''(s)} = -\frac{as^2 + bs + c}{s^2 + ds + \omega_n^2} , \tag{3.162}$$

where coefficients a, b, c, d, and ω_n contain the system parameters. When $d = 0$, the flow oscillates with frequency ω_n. For $d > 0$, the oscillations introduced by a disturbance in heat flux vanish for large values of time. Quandt[158] suggested that $d = 0$ be used as a stability criterion, but the validity of this procedure is limited to small oscillations due to the linearization approximation.

The linearized STABLE-3 program[159] has been compared with existing data[148,160-163] by Neal and Zivi,[149] who found that STABLE-3 predicts data within 20% for ~70% of the tests. Jain[164] recently showed that STABLE-3 predicts data within 20% for 90% of his experiments. However, one limitation of the STABLE-3 program is that it requires prior knowledge of the steady-state flow rate for a given power level.

Nonlinear approaches to this problem were considered by Randles,[165] Garlid,[166] and Nahavandi and Von Hollen.[167] The last investigators analyzed the stability of U-tube steam generators considering the dimensions of length along the flow path and time. Differential conservation equations were solved numerically by a "Modified-Euler" integration process combined with appropriate iteration procedures. Nahavandi and Von Hollen believe this procedure avoids the approximations inherent in linearized techniques.

3-3.3.3 Simplified Criterion for Dynamic Instability

It is desirable to be able to obtain a preliminary estimate of the dynamic stability boundary under many circumstances. If the boiling channel can be considered to be placed between two plenums across which a constant pressure drop is maintained (equivalent to a feed pump with a flat characteristic in the range of interest), Ishii's[168] stability criterion can be used. At high inlet subcooling, he concluded that the system is stable when

$$N_{pch} \leqslant N_{sub} + \frac{2 [K_i + (f_m/2D_h^*) + K_e]}{1 + \frac{1}{2} [(f_m/2D_h^*) + 2K_e]} , \tag{3.163}$$

where

$\quad N_{pch}$ = equilibrium phase change number = $\Omega_{eq} L/V_i$

$\quad \Omega_{eq}$ = equilibrium frequency of phase change, rad/s

$\quad\quad = \Gamma_g \Delta\rho/(\rho_g \rho_l)$

$\quad \Gamma_g$ = mass rate of vapor generation per unit vol, mass/vol s

$\quad N_{sub}$ = subcooling number = $(\Delta\rho/\rho_g)(\Delta H_{sub}/\Delta H_{fg})$

$\quad \Delta H_{sub}$ = inlet subcooling = $H_{sat} - H_{in}$

ΔH_{fg} = enthalpy change on evaporation

H_{in}, H_{sat} = enthalpy of liquid at inlet and saturation, respectively

ρ_g, ρ_l = gas and liquid densities, respectively

$\Delta\rho = \rho_l - \rho_g$

f_m = two-phase mixture friction factor

D_h = hydraulic diameter, $D_h^* = D_h/L$

L = length of heated channel

K_i, K_e = inlet and exit orifice coefficient, $\Delta P = KG^2/\rho g_c$

V_i = fluid velocity at channel inlet.

The stability criterion of the previous equation holds only if $N_{sub} > (N_{sub})_{cr}$ and $(N_{sub})_{cr}$ is obtained from[169]

$$(N_{sub})_{cr} = 0.0022 \ (Pe)(A_c/p_hL)(N_{pch})_0 \qquad (3.164)$$

$$(N_{sub})_{cr} = 154 \ (A_c/p_hL)(N_{pch})_0 \quad \text{if} \quad Pe \geqslant 70\ 000 \ , \qquad (3.165)$$

where

Pe = Peclet number for liquid at inlet conditions

A_c = cross-sectional area of boiling channel

p_h = heated perimeter of boiling channel

$(N_{pch})_0$ = value of N_{pch} determined from linearized stability calculations for zero subcooling at the inlet.

Under most conditions, $(N_{pch})_0$ occurs at channel exit quality x_e, which is close to 1.0. Therefore, an approximate value for $(N_{pch})_0$ can be obtained from this fact and the relationship

$$(N_{pch}) = x_e(\Delta\rho/\rho_g) \ . \qquad (3.166)$$

At inlet subcooling numbers below $(N_{sub})_{cr}$, the system is expected to be stable if

$$(N_{pch}) \leqslant \left[\frac{(N_{pch})_{cr} - (N_{pch})_0}{(N_{sub})_{cr}}\right] N_{sub} + (N_{pch})_0 \ , \qquad (3.167)$$

where $(N_{pch})_{cr}$ = the value of N_{pch} obtained from Eq. (3.163) when $N_{sub} = (N_{sub})_{cr}$. The Freon loop stability data of Saha et al.[169] appear to be in good agreement with this simplified approach.

3-3.3.4 Momentum Integration

This procedure is useful in treating the problem of parallel channel instability. Nonlinear transient momentum equations can be conveniently solved for several channels by integrating the momentum equation along each channel. Fluid properties are obtained from the energy equation and the equation-of-state; integrated equations are then solved simultaneously. These techniques have been incorporated into the HYDNA computer code,[170] which predicts flow oscillation in parallel channels. To describe the HYDNA computational method, we again follow Tong.[23]

It is useful to separate inlet mass velocity and the instantaneous axial mass flow rate distribution. If $\Phi(z,t)$ is the axial mass flow rate distribution in the channel and $G(0,t)$ is the inlet mass flow rate, which can vary with time, we have

$$G(z,t) = G(0,t)\Phi(z,t) \ . \tag{3.168}$$

Substituting this, where $G(z,t)$ is mass velocity as a function of z and t, into the momentum balance over the channel length gives

$$\frac{dG(0,t)}{dt} = - \frac{\Delta P + \int_0^L Fdz + \int_0^L \pm \rho dz + \frac{1}{g_c}\Delta(v'G^2) + \frac{1}{g_c}G(0,t)\int_0^L \frac{\partial\Phi}{\partial t}dz}{\frac{1}{g_c}\int_0^L \Phi dz} \ ,$$

$$\tag{3.169}$$

where ΔP is the pressure difference between the common plenums and $v' = 1/\rho$ for homogeneous flow. Factor $\int_0^L \Phi dz$, which accounts for the effect of axial variation of the mass velocity, comes close to unity for a slow transient if G remains nearly constant throughout the passage. However, $\partial\Phi/\partial t$ vanishes if the shape of $G(z)$ remains constant. All terms of Eq. (3.169), except $dG(0,t)/dt$ and ΔP, can be easily determined. Let C and B be given as

$$C = - \frac{1}{\frac{1}{g_c}\int_0^L \Phi dz} \ , \tag{3.170}$$

and

$$B = - \frac{\int_0^L Fdz + \int_0^L \pm \rho dz + \frac{1}{g_c}\Delta(v'G^2) + \frac{1}{g_c}G(0,t)\int_0^L \frac{\partial\Phi}{\partial t}dz}{\frac{1}{g_c}\int_0^L \Phi dz} \ . \tag{3.171}$$

Substitution into the momentum balance yields

$$\frac{dG(0,t)}{dt} = C\Delta P + B \ . \tag{3.172}$$

Equation (3.172) can be written for each of n parallel channels

$$\frac{dG_1(0,t)}{dt} = C_1 \Delta P + B_1$$

$$\frac{dG_2(0,t)}{dt} = C_2 \Delta P + B_2$$

$$\cdots\cdots\cdots\cdots$$

$$\frac{dG_n(0,t)}{dt} = C_n \Delta P + B_n \ . \tag{3.173}$$

With a given total inlet flow rate $W(t)$,

$$A_1 G_1(0,t) + A_2 G_2(0,t) + \ldots + A_n G_n(0,t) = W(t) \ , \tag{3.174}$$

where $A_1, A_2, \ldots, A_n$ are the respective flow areas of channels $1, 2, \ldots, n$. Differentiation of Eq. (3.174) gives

$$A_1 \frac{dG_1(0,t)}{dt} + A_2 \frac{dG_2(0,t)}{dt} + \ldots + A_n \frac{dG_n(0,t)}{dt} = \frac{dW}{dt} \ ,$$

or

$$A_1(C_1 \Delta P + B_1) + \ldots + A_n(C_n \Delta P + B_n) = \frac{dW}{dt} \ . \tag{3.175}$$

The latter equation can be used to evaluate the pressure drop that is common to all channels; i.e.,

$$\Delta P = \frac{(dW/dt) - \Sigma B_n A_n}{\Sigma C_n A_n} \ . \tag{3.176}$$

Equation (3.175) can be used to predict the pressure drop as a function of time by iterating on it along with n simultaneous equations of Eq. (3.173). The steps in the calculation for a small time increment are given by Tong[23] with a method that can predict the thermal and hydraulic behavior of the system in response to a specified variable inlet flow rate or a given variable power input to the channel.

Parallel channel instability can also be predicted using approaches other than momentum integration. Moxon's[150] LOOP 20 code is an example of one such alternative.

3-3.4 Recommended Methods of Analysis

The recommended methods of analysis for various types of flow instabilities are:

1. The onset of density wave or dynamic instability in a uniformly heated boiling channel can be predicted by a linearization of the basic equations. Benchmarking such procedures against experimental data or more complex nonlinear procedures is desirable.

2. For a parallel boiling channel with nonuniform heat flux distributions, a momentum integration approach such as that used in the HYDNA code[170] is convenient.

3. The static stability of a natural or forced circulation loop can be determined by the Ledinegg criterion[147] with a lumped-pressure drop.

4. The analysis for predicting the onset of flow pattern instability has not been well established. An experimental determination is recommended.

5. To predict the onset of thermal oscillation, empirical correlations[151] are suggested.

3-4 LIQUID ENTRAINMENT, VAPOR CARRY-UNDER, AND STEAM SEPARATION

Liquid entrainment during evaporation, vapor carry-under in a natural circulation loop, and steam separation are closely related because they are all controlled by buoyancy forces. Although they are usually considered to be in the province of the boiling water reactor, they must be understood for steam generator design and system behavior during a LOCA.

3-4.1 Entrainment, Carry-Over, and Flooding

3-4.1.1 Liquid Injection

Liquid droplets injected into a gas stream by the burst of bubbles have been studied by Mitsuishi et al.,[171] who found:

1. The liquid drop spray is injected almost perpendicularly to the liquid surface.

2. The liquid injected is largely from bubbles with diameters of 2 mm or less, and

$$D_p \propto D_b^{3/2} \ , \tag{3.177}$$

where D_p is the diameter of the droplet and D_b is the diameter of the bubble.

3. Variation in the surface tension of the liquid has little effect on drop size. An increase in the viscosity of liquid brings about a decrease in drop diameter and lower jetting height.

Injection velocities and jetting heights obtained by Mitsuishi et al.[171] are shown in Figs. 3-23 and 3-24, respectively. They also found that liquid entrainment increased rapidly after the vapor mass velocity increased above 800 lb/(h ft²) (~6 ft/s) at atmospheric pressure.

Newitt et al.[172] reported that generation of drops by bursting bubbles is influenced by the depth of bubble generation, by bubble diameter, and by

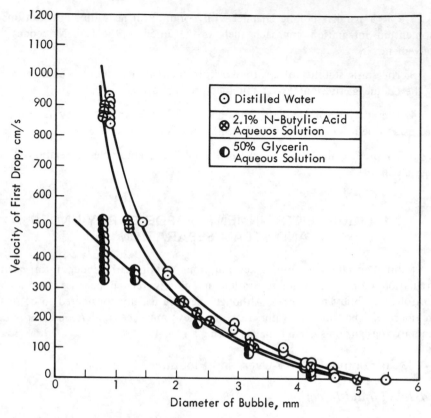

Fig. 3-23.   Velocity of first drop versus diameter of bubble (from Ref. 171).

physical properties of the liquid. They found that large liquid drops can be estimated by increasing the vapor space in an evaporator or by increasing plate spacing in a plate column. The number of large drops is usually few for bubbles generated at very shallow depths. Sterman[173] provides an expression for determining minimum diameter D at which entrainment is independent of vessel size:

$$\frac{D[\rho_v/(\rho_l - \rho_v)]^{-0.2}}{[\sigma/(\rho_l - \rho_v)]^{1/2}} \gtrsim 260 \ , \tag{3.178}$$

where

σ = surface tension, lb/ft

ρ_1 = liquid density, lb/ft^3

ρ_v = vapor density, lb/ft^3

D = diameter, ft.

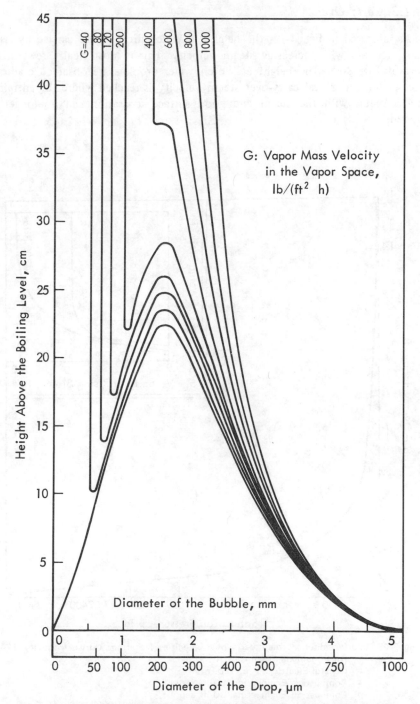

Fig. 3-24.   Height of drop entrained above the boiling liquid level (from Ref. 171).

3-4.1.2 Carry-Over

As indicated in Fig. 3-24, the height to which droplets are carried by rising steam increases with increasing steam velocity. This is most clearly seen in the insert of Fig. 3-25. The height at which water droplets are observed gradually increases until a critical carry-over steam velocity is reached where the entrained liquid remains with the steam and good surface separation can no longer be achieved.

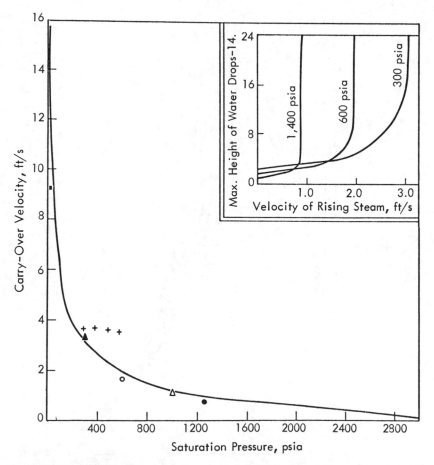

Fig. 3-25.   Surface disengagement carry-over velocities. Notation: Curves from Ref. 174.
+   Allis Chalmers, 33-in. sep. ht.
△   General Electric, 32-in., sep. ht.
▲   Combustion Engineering, 50-in. sep. ht.
○   Combustion Engineering, 30-in. sep. ht.
●   Combustion Engineering, 24-in. sep. ht.
■   From Ref. 171.

Carry-over at high pressures has been studied by Davis[174] and various investigators on AEC-sponsored programs. Carry-over velocity as a function of pressure is plotted in Fig. 3-25, where the insert shows that at high pressures, carry-over occurs much more suddenly and at a much lower steam velocity. Wilson and McDermott's[175] data can be used in evaluating carry-over in small flow channels with cold walls.

Significant carry-over would be encountered during the reflood phase of a hypothetical LOCA. In this period, the core would be partly refilled, but there would be substantial steam generation due to the decay heat released by the fuel rods. When the average steam velocity exceeds the carry-over velocity (Fig. 3-25), a portion of the water being added to reflood the core is removed by entrainment. The amount of this entrainment is generally estimated from empirical correlations of test data. The rate of entrainment, $\dot{M}_E$, is sometimes given in terms of the carry-over fraction. That is,

$$\dot{M}_E = (CRF)(\dot{M}_f) - \dot{M}_{sg} \ , \tag{3.179}$$

where

$\dot{M}_f$ = rate at which water is added to core, mass/time

CRF = carry-over fraction

$\dot{M}_{sg}$ = rate of steam generation in the core, mass/time.

The carry-over fraction (rate of total core exit flow rate to core inlet flow rate) was correlated by[176]

$$CRF = 0.9553 \exp[-0.0013935(P/60)^2][1 - \exp(-4.3127(q'/1.24)]$$

$$\times \{1 - \exp[-16.62(V/6)]\} \exp[-0.037161(\Delta T/140)^2]$$

$$\times \frac{1 + \sin 3.1416[-0.5 + (Z/6)^{0.025733(6/V)+0.13553(\Delta T/140)+0.0656(P/60)}]}{2} \ ,$$

$$\tag{3.180}$$

where

P = pressure, psia

q' = peak linear heat rate, kW/ft

V = flooding rate, in./s

ΔT = coolant subcooling at inlet, °F

Z = quench front level, ft.

Note that Eq. (3.180) is an empirical correlation of data for PWR ECCS evaluation during a hypothetical LOCA. It should not be used for other conditions.

3-4.1.3 Flooding

The flooding phenomenon, which can be considered somewhat related to the subject of carry-over, arises during the latter stages of blowdown following a large cold-leg break in a hypothesized LOCA. Steam generated in the core flows up the annulus between the core barrel and the pressure vessel on its way to the break. Simultaneously, emergency core cooling water flows toward the same region and attempts to flow toward the lower plenum. It is desired to refill the plenum region and then flood the core. When high upward steam velocities are reached in the annulus, part or all of the emergency cooling water is prevented from getting to the lower plenum. Excess water exits through the break with the steam. This limitation on the rate of liquid penetration is imposed by "flooding," which is the term generally used to describe the boundary of allowable liquid-gas counter-current flow rates.

Small-scale flooding tests with both air-water and steam-water systems have been performed at several installations.[177-179] In general, correlations of observed behavior have been in terms of dimensionless groups j_f^* and j_g^*, where

$$j_f^* = (Q_l/A)\rho_l^{1/2}[Lg(\rho_l - \rho_g)]^{-(1/2)} \tag{3.181}$$

$$j_g^* = (Q_g/A)\rho_g^{1/2}[Lg(\rho_l - \rho_g)]^{-(1/2)} \, , \tag{3.182}$$

where

A = cross-sectional flow area

Q_g, Q_l = water and vapor volumetric flow rates, respectively

ρ_g, ρ_l = gas and liquid densities, respectively

g = acceleration of gravity

L = characteristic dimension.

On the basis of tests with saturated cooling water, Block et al.[180] correlated the limiting liquid flow rate by

$$(j_g^*)^{1/2} + m(j_{f\text{down}}^*)^{1/2} = C \, , \tag{3.183}$$

where

$C = 0.32$

$m = \exp[-5.6(j_{f\text{in}}^*)^{0.6}]$

$j_{f\text{in}}^*$ = dimensionless group of Eq. (3.181) based on supplied liquid flow and defined with L = total gap circumference

$j_{f\text{down}}^*$ = dimensionless group of Eq. (3.181), which indicates fluid actually reaching lower plenum and defined with L = total gap circumference.

Note that experimental steam-water data do not show a dependence on gap thickness. Use of total gap circumference as the characteristic L in the j^* groups is somewhat arbitrary and may not be correct.

When subcooled water is used as the injection fluid, a portion of the upward steam flow is condensed. This reduces steam velocity and allows increased liquid penetration. Block et al.[180] suggest that a modified dimensionless steam flux, $j^*_{g\text{mod}}$, be computed from

$$j^*_{g\text{mod}} = j^*_g - [F(\Delta T)_{\text{sub}}(c_P/H_{fg})(\rho_l/\rho_g)^{1/2} j^*_{f\text{in}}] , \qquad (3.184)$$

where

c_P = specific heat of liquid

ΔT_{sub} = saturation temperature less actual coolant temperature

$F = (P)^{1/4}(1 + b j^*_{f\text{in}})^{-1}$

P = pressure, atm

b = 16 for flat plate geometry, 30 for cylindrical geometry.

The effects of various geometry changes are still (1977) imperfectly understood. Short baffles or guide vanes inserted into the annulus seem to allow improved water delivery at high values of $j f^*_{f\text{in}}$. Long baffles appear to lead to a further improvement in liquid delivery. The effect of the number and location of hot- and cold-leg pipes in the annulus is under study (1977).

Another variable that must be considered is the temperature of the core and vessel walls.[181] These walls can be superheated during the latter stages of blowdown and therefore lead to additional steam generation. It is total steam flow that causes flooding and limits water penetration.[d]

3-4.2 Carry-Under

Carry-under, the entrainment of bubbles of vapor in the liquid stream leaving a separating surface, is the inverse of carry-over. Carry-under in both steam-water and air-water separations was studied by Petrick,[39] whose air-water data were obtained using riser radii of 3.25 and 5.5 in., respectively, and whose steam-water data were taken at pressures of 600, 1000, and 1500 psia. He also obtained data from the Experimental Boiling Water Reactor (EBWR) at 300 and 600 psia and suggested two carry-under equations:

For X_D/X_R = 3 to 64,

$$\frac{X_D}{X_R} = -0.04 \log \left[\frac{\left(\dfrac{V_g}{V_e}\right)\left(\dfrac{\sigma^{2/3}}{G^2\mu}\right)\left(\dfrac{\rho_l}{\rho_v}\right)^{1/2}\left(\sqrt{\dfrac{D}{Z}} + \sqrt{\dfrac{Z}{D}}\right)}{64} \right] . \qquad (3.185)$$

---

[d]The reader who is unfamiliar with the ECCS provided for a hypothetical LOCA is referred to Sec. 5-4 for a description of the system.

For $\chi_D/\chi_R = 0.1$ to 3,

$$\frac{\chi_D}{\chi_R} = -0.6 \log \left[\frac{\left(\frac{V_g}{V_e}\right)\left(\frac{\sigma^{2/3}}{G^2\mu}\right)\left(\frac{\rho_l}{\rho_v}\right)^{1/2}\left(\sqrt{\frac{D}{Z}} + \sqrt{\frac{Z}{D}}\right)}{3.7} \right] , \qquad (3.186)$$

where

χ_D = steam quality in the downcomer

χ_R = steam quality in the riser

V_g = vapor velocity in the riser, ft/s

V_e = velocity of entrained gas in the downcomer, ft/s

D = riser diameter

Z = interface height

σ = surface tension, lb/ft

μ = viscosity of liquid, lb/(s ft)

G = liquid mass velocity in the downcomer, lb/(s ft^2).

Petrick[39] also suggested a correlation for calculating the slip ratio in downflow

$$\frac{V_v}{V_l} = 0.63 \left(\frac{\bar{V}^2}{gD}\right)^{0.4} \left(\frac{\chi}{1-\chi}\frac{\rho_l}{\rho_v}\right)^{0.2} , \qquad (3.187)$$

where $\bar{V}$ is average velocity of the mixture.

3-4.3 Steam Separation

3-4.3.1 Separation by Surface Disengagement

Surface disengagement is effective when the steam velocity is low enough for water droplets to return to the two-phase interface by gravitational force. In designing this type of steam separator, the data in Fig. 3-25 can be used. The increase in vapor density with pressure initially overcomes the decrease in escape velocity, so the allowable steam release rate reaches maximum at 1750 psia. This allowable steam release rate determines the maximum power output of a boiler relying solely on surface disengagement for its steam separation.

3-4.3.2 Separation by Centrifugal Force

This mechanism is encountered in a radial- or vane-type steam separator. For analysis in the range of interest, the effects of gravity force, slip, and turbulence can be neglected, and the dynamic equation of a bubble becomes

$$\frac{C_d \pi D_b^2}{4} \rho_l \frac{u^2}{2g_c} = \frac{1}{6} \pi D_b^2 \frac{(\rho_l - \rho_v)}{g_c} \frac{V^2}{R} , \qquad (3.188)$$

where

C_d = drag coefficient, dimensionless

D_b = bubble diameter, ft

R = radius of separator, ft

u = radial velocity, ft/s

V = tangential velocity, ft/s.

By rearranging the above equation,

$$u^2 = \frac{4V^2 D_b}{3RC_d} .$$ (3.189)

For a vane separator, design equations can be written by equating the time required for the bubble to separate

$$\frac{R\theta}{V} = \frac{b}{u} ,$$ (3.190)

where b is the water-annulus thickness that the bubble travels and θ is the angle the bubble travels along the vane before separation. If we combine Eqs. (3.189) and (3.190), we get the condition of bubble separation[182]:

$$\frac{b^2}{R\theta} = \frac{4D_b}{3C_d} .$$ (3.191)

From Eq. (3.190), we see that maximum inlet velocity V (i.e., maximum mixture rate) can be increased by increasing R and by decreasing b. Vane arc θ should be large, but not more than 130 to 180 deg to allow space for the separated water to leave the vane.

3-5 SUMMARY

Analyses of fluid system behavior must consider both steady-state and transient operation. A proper characterization of steady-state behavior should include component pressure losses, flow redistribution, flow mixing, and stability to flow perturbations. Further, since substantial boiling occurs in the hottest channels of a PWR core, the effects of two-phase flow on the foregoing must be taken into account.

Under hypothetical LOCA conditions, the bulk of the system remains subcooled very briefly. During this brief period, rarefaction and compression waves passing through the system can lead to significant pressure differentials across components. These differentials must be determined. In the subsequent portions of the accident sequence, two-phase effects predominate. Critical flow at the break determines the rate at which fluid escapes and, hence, the rate of system depressurization. Calculations of pressure drops around the system must include two-phase friction and acceleration effects. In the latter stages of the accident, carry-over and flooding phenomena have significant roles in determining the rate at which the core is reflooded.

REFERENCES

1. L. F. Moody, "Friction Factors for Pipe Flow," *Trans. ASME*, **66**, 671 (1944).

2. W. M. Kays and A. L. London, *Compact Heat Exchangers*, The National Press, Palo Alto, California (1955).

3. V. L. Streeter, *Handbook of Fluid Dynamics*, John Wiley & Sons, Inc., Publishers, New York (1961).

4. S. Levy, "Fluid Flow," from *Technology of Nuclear Reactor Safety*, Vol. II, Chap. 15, by T. J. Thompson and J. G. Beckerly, MIT Press, Cambridge, Massachusetts (1973).

5. I. E. Idelchik, *Handbook of Hydraulic Resistance*, Israel Program for Scientific Translation (1968).

6. E. N. Sieder and G. E. Tate, "Heat Transfer and Pressure Drop of Liquid in Tubes," *Ind. Eng. Chem.*, **28**, 1429 (1936).

7. W. H. Esselman, I. H. Mandil, S. J. Green, P. C. Ostergaard, and R. A. Frederickson, "Thermal and Hydraulic Experiments for Pressurized Water Reactors," *Proc. 2nd U. N. Int. Conf. Peaceful Uses Atomic Energy*, 7, 758 (1958).

8. B. W. LeTourneau, R. E. Grimble, and J. E. Zerbe, "Pressure Drop for Parallel Flow Through Rod Bundles," *Trans. ASME*, **79**, 483 (1957).

9. O. J. Mendler, A. S. Rathbun, N. E. Van Huff, and A. Weiss, "Natural Circulation Tests with Water at 800 to 2000 psia Under Non-Boiling, Local Boiling, and Bulk Boiling Conditions," *Trans. ASME, Ser. C, J. Heat Transfer*, **83**, 261 (1961).

10. W. M. Rohsenow and J. A. Clark, "Heat Transfer and Pressure Drop Data for High Heat Flux Densities to Water at High Sub-Critical Pressures," *Proc. 1951 Heat Transfer and Fluid Mechanics Institute*, Stanford University Press, Stanford, California (1951).

11. R. G. Deissler and M. F. Taylor, "Analysis of Axial Turbulent Flow and Heat Transfer Through Banks of Rods or Tubes," *Proc. Reactor Heat Transfer Conf.*, Part 1, Book 2, p. 416, TID-7529, U.S. Atomic Energy Commission, Washington, D. C. (1957).

12. R. G. Deissler, "Analysis of Turbulent Heat Transfer, Mass Transfer, and Friction in Smooth Tubes at High Prandtl and Schmidt Numbers," NACA-1210, National Advisory Committee for Aeronautics (1955).

13. R. G. Deissler, "Analytical and Experimental Investigation of Adiabatic Turbulent Flow in Smooth Tubes," NACA-TN-2138, National Advisory Committee for Aeronautics (1953).

14. J. Laufer, "The Structure of Turbulence in Fully Developed Pipe Flow," NACA-TN-2954, National Advisory Committee for Aeronautics (1953).

15. P. Miller, J. J. Byrnes, and D. M. Benforado, "Heat Transfer to Water Flowing Parallel to a Rod Bundle," *AIChE J.*, **2**, 226 (1956).

16. J. L. Wantland, "Compact Tubular Heat Exchangers," *Proc. Reactor Heat Transfer Conf.*, Part 1, Book 2, p. 525, TID-7529, U. S. Atomic Energy Commission, Washington, D. C. (1957).

17. D. A. Dingee and J. W. Chastain, "Heat Transfer from Parallel Rods in Axial Flow," *Proc. Reactor Heat Transfer Conf.*, Part 1, Book 2, p. 462, TID-7529, U. S. Atomic Energy Commission, Washington, D. C. (1957).

18. R. G. Deissler and M. F. Taylor, "Analysis of Turbulent Flow and Heat Transfer in Non-Circular Passages," NACA-TN-4384, National Advisory Committee for Aeronautics (1958).

19. A. C. Trupp and R. S. Azad, "Structure of Turbulent Flow in Triangular Rod Bundles," *Nucl. Eng. Des.*, **32**, 47 (1975).

20. R. T. Berringer and A. A. Bishop, "Model of the Pressure Drop Relationships in a Typical Fuel Rod Assembly," YAEC-75, Westinghouse Atomic Power Division (1959).

21. A. N. De Stordeur, "Drag Coefficients for Fuel-Element Spacers," *Nucleonics*, **19**, *6*, 74 (1961).

22. K. Rehme, "Pressure Drop Correlations for Fuel Element Spacers," *Nucl. Technol.*, **17**, 15 (1973).

23. L. S. Tong, *Boiling Heat Transfer and Two-Phase Flow*, John Wiley & Sons, Inc., Publishers, New York (1964).

24. G. W. Maurer, "A Method of Predicting Steady-State Boiling Vapor Fraction in Reactor Coolant Channels," *Bettis Technical Review*, WAPD-BT-19, p. 59, Bettis Atomic Power Laboratory (1960).

25. R. W. Bowring, "Physical Model, Based on Bubble Detachment, and Calculation of Steam Voidage in the Subcooled Region of a Hot Channel," HPR-29, OECD Halden Reactor Project, Halden, Norway (1962).

26. D. Maitra and S. Raju, "Vapor Void Fraction in Subcooled Boiling," *Nucl. Eng. Des.*, **32**, 20 (1975).

27. P. Griffith, J. A. Clark, and W. M. Rohsenow, "Void Volumes in Subcooled Boiling Systems," 58-HT-19, American Society of Mechanical Engineers, New York (1958).

28. F. W. Staub, "Void Fraction in Subcooled Boiling—Prediction of Initial Point of Net Vapor Generation," *Trans. ASME Series C, J. Heat Transfer*, **90**, 151 (1968).

29. S. Levy, "Forced Convection Subcooled Boiling Prediction of Vapor Volumetric Fraction," *Int. J. Heat Transfer*, **10**, 951 (1967).

30. J. R. S. Thom, W. M. Walker, T. A. Fallow, and G. F. S. Reising, "Boiling in Subcooled Water During Flow Up Heated Tubes or Annuli," *Proc. Inst. ME, 1965-1966*, Part 3C, p. 180 (1966).

31. L. S. Tong, "Two-Phase Flow in Nuclear Reactors, and Flow Boiling and Its Crises," Lecture Notes for a University of Michigan Short Course (Summer 1968).

32. S. G. Bankoff, "A Variable Density Single-Fluid Model for Two-Phase Flow with Particular Reference to Steam-Water Flow," *Trans. ASME, Ser. C, J. Heat Transfer*, **82**, 265 (1960).

33. G. E. Kholodovski, "New Method for Correlating Experimental Data for the Flow of a Steam-Water Mixture in Vertical Pipes," *Teploenergetica*, **4**, *7*, 68 (1957).

34. G. A. Hughmark, "Holdup in Gas-Liquid Flow," *Chem. Eng. Prog.*, **58**, *4*, 62 (1962).

35. N. Zuber and J. A. Findlay, "Average Volumetric Concentration in Two-Phase Flow System," *Trans. ASME, J. Heat Transfer*, **87**, 453 (1965).

36. P. Griffith and G. B. Wallis, "Two-Phase Slug Flow," *Trans. ASME, Ser. C, J. Heat Transfer*, **83**, 307 (1961).

37. T. Harmathy, "Velocity of Large Drops and Bubbles in Media of Infinite and of Restricted Extent," *AIChE J.*, **6**, 281 (1960).

38. J. F. Marchaterre and B. M. Hoglund, "Correlation for Two-Phase Flow," *Nucleonics*, **20**, *8*, 142 (1962).

39. M. Petrick, "A Study of Carry-Under Phenomena in Vapor Liquid Separation," *AIChE J.*, **9**, *2*, 253 (1963).

40. R. C. Martinelli and D. B. Nelson, "Prediction of Pressure Drops During Forced Circulation Boiling of Water," *Trans. ASME*, **70**, 695 (1948).

41. J. R. S. Thom, "Prediction of Pressure Drop During Forced Circulation Boiling of Water," *Int. J. Heat Mass Transfer*, **7**, 709 (1964).

42. W. L. Owens and V. E. Schrock, "Local Pressure Gradients for Subcooled Boiling of Water in Vertical Tubes," 60-WA-249, American Society of Mechanical Engineers, New York (1960).

43. B. W. Le Tourneau and M. Troy, "Heating, Local Boiling, and Two-Phase Pressure Drop for Vertical Upflow of Water at Pressures Below 1850 psia," WAPD-TH-410, Bettis Atomic Power Laboratory (1960).

44. N. V. Tarasova et al., "Pressure Drop of Boiling Subcooled Water and Steam-Water Mixture Flow in Heated Channels," *Proc. 3rd Int. Heat Transfer Conf.*, pp. 178-183, American Society of Mechanical Engineers, New York (1966).

45. W. L. Owens, Jr., "Two-Phase Pressure Gradients," *Int. Developments Heat Transfer*, Part II, pp. 363-368, American Society of Mechanical Engineers, New York (1961).

46. W. H. McAdams et al., "Vaporization Inside Horizontal Tubes, II—Benzene-Oil Mixtures," *Trans. ASME*, **64**, 193 (1942).

47. W. G. Choe, "Flow Patterns and Pressure Drop in Horizontal Two-Phase Flow," PhD Thesis, University of Cincinnati (1975).

48. C. J. Baroczy, "A Systematic Correlation for Two-Phase Pressure Drop," NAA-SR-Memo-11858, North American Aviation (Mar. 1966).

49. R. W. Lockhart and R. C. Martinelli, "Proposed Correlation of Data for Isothermal Two-Phase, Two-Component Flow in Pipes," *Chem. Eng. Prog.*, **45**, 39 (1949).

50. P. A. Lottes, "Expansion Losses in Two-Phase Flow," *Nucl. Sci. Eng.*, **9**, 26 (1961).

51. E. Janssen, "Two-Phase Pressure Loss Across Abrupt Contractions and Expansions—Steam-Water at 600 to 1400 psia," *Proc. Int. Heat Transfer Conf.*, **5**, 13 (1966).

52. J. Weisman, A. Husain, and B. Harshe, "Two-Phase Pressure Drop Across Abrupt Changes and Restriction," in *Two-Phase Transport and Reactor Safety*, T. N. Veziroglu and S. Kakac, Eds., Hemisphere Publishing Company, Washington, D. C. (1978).

53. D. Chisholm, "Pressure Losses in Bends and Tees During Steam-Water Flow," Report No. 318, U.K. National Engineering Laboratory (1967).

54. D. Chisholm and L. A. Sutherland, "Prediction of Pressure Gradients in Pipeline Systems During Two-Phase Flow," Paper No. 4, Symp. Fluid Mechanics and Measurements in Two-Phase Flow Systems, Leeds, England (1967).

55. J. Weisman, A. H. Wenzel, L. S. Tong, D. Fitzsimmons, W. Thorne, and J. Batch, "Experimental Determination of the Departure from Nucleate Boiling in Large Rod Bundles at High Pressures," *Chem. Eng. Prog. Symp. Ser.*, **64**, *82*, 114 (1968).

56. D. S. Rowe, "Cross-Flow Mixing Between Parallel Flow Channels During Boiling, Part I," BNWL-371P71, Pacific Northwest Laboratory (Mar. 1967).

57. W. H. Bell and B. W. Le Tourneau, "Experimental Measurements of Mixing in Parallel Flow Rod Bundles," WAPD-TH-381, Bettis Atomic Power Laboratory (1960).

58. R. J. Sembler, "Mixing in Rectangular Nuclear Reactor Channels," WAPD-T-653, Bettis Atomic Power Laboratory (1960).

59. E. D. Waters, "Fluid Mixing Experiments with a Wire-Wrapper 7-Rod Bundle Fuel Assembly," HW-70178 Rev., Hanford Laboratory, General Electric Company (1963).

60. D. S. Rowe and C. W. Angle, "Cross-Flow Mixing Between Parallel Flow Channels During Boiling, Part II," BNWL-371, Pacific Northwest Laboratory (1967).

61. J. T. Rodgers and R. G. Rosehart, "Turbulent Interchange Mixing in Fuel Bundles," Trans. Conf. Applied Mech., University of Waterloo, Waterloo, Canada (1969).

62. J. T. Rodgers and R. G. Rosehart, "Mixing by Turbulent Interchange Bundles, Correlation and Inference," Paper 72-H7-53, ASME-AIChE Heat Transfer Conf., American Society of Mechanical Engineers, New York (1972).

63. K. Petrunik, "Turbulent Interchange in Simulated Rod Bundles," PhD Thesis, University of Windsor, Canada (1973).

64. K. Singh, "Air-Water Turbulent Mixing in Simplified Rod Bundle Geometries," PhD Thesis, University of Windsor, Canada (1972).

65. T. Van der Ros and M. Bogaart, Nucl. Eng. Des., 12, 259 (1970).

66. K. P. Galbraith and J. G. Knudson, Chem. Eng. Prog. Symp. Ser., 68, 118, 90 (1972).

67. D. S. Rowe and C. W. Angle, "Cross-Flow Mixing Between Parallel Flow Channels During Boiling, Part III, Effect of Spacers on Mixing Between Channels," BNWL-371, Pacific Northwest Laboratory (1969).

68. K. F. Rudzinski, K. Singh, and C. C. St. Pierre, Can. J. Chem. Eng., 50, 297 (1972).

69. J. M. Gonzalez-Santalo and P. Griffith, "Two-Phase Flow Mixing in Rod Bundle Subchannels," Paper 72-WA/NE-19, ASME Winter Mtg., American Society of Mechanical Engineers, New York (1972).

70. J. L. du Bousquet and J. Bouré, "Melange par Diffusion Turbulence Entre Deux Sous-Canaux d'en Reacteur à Grappe," European Two-Phase Flow Group Mtg., Karlsruhe, West Germany (1969).

71. K. Singh and C. C. St. Pierre, Nucl. Sci. Eng., 50, 382 (1973).

72. S. G. Beus, "A Two-Phase Turbulent Mixing Model for Flow in Rod Bundles," WAPD-2438, Bettis Atomic Power Laboratory (1971).

73. R. Lahey, Jr. and F. A. Schraub, "Mixing Flow Regimes and Void Fraction for Two-Phase Flow in Rod Bundles," in Two-Phase Flow and Heat Transfer in Rod Bundles, V. E. Schrock, Ed., p. 1, American Society of Mechanical Engineers, New York (1969).

74. H. Chelemer, J. Weisman, and L. S. Tong, "Subchannel Thermal Analysis of Rod Bundle Cores," Nucl. Eng. Des., 21, 3 (1972).

75. R. K. Kim and R. H. Stoudt, "Cross Flow Between Rod Bundles with Spacer Grids," Trans. Am. Nucl. Soc., 22, 584 (1975).

76. J. Weisman, "Cross Flow Resistance in Rod Bundle Cores," *Nucl. Technol.*, 15, 465 (1971).

77. D. S. Rowe, "COBRA-IIIC, A Digital Computer Program for Steady State and Transient Thermal Analysis of Rod Bundle Nuclear Fuel Elements," BNWL-1695, Battelle Northwest Laboratory (1973).

78. Z. Rouhani, "Axial and Transverse Momentum Balance in Subchannel Analyses," Topl. Mtg. Requirements and Status of the Prediction of the Physics Parameters for Thermal and Fast Reactors, Jülich, West Germany (1973).

79. W. D. Brown, E. U. Khan, and N. E. Todreas, *Nucl. Sci. Eng.*, 57, 164 (1975).

80. W. T. Sha, A. A. Szewczyk, R. C. Schmitt, T. H. Hughes, and P. R. Huebotter, "Cross-Flow Approximations Used in the Thermal-Hydraulic Multichannel Analysis," *Trans. Am. Nucl. Soc.*, 18, 134 (1974).

81. J. M. Madden, "Two-Phase Air-Water Flow in a Slot Type Distributor," MS Thesis, University of Windsor, Canada (1968).

82. J. Weisman, "Review of Two-Phase Mixing and Diversion Cross Flow in Subchannel Analyses," AEEW-R-928, U. K. Atomic Energy Authority, Winfrith, England (1973).

83. G. L. Shires and G. Hewitt, Personal Communication, U. K. Atomic Energy Authority, Harwell, England (1973).

84. M. A. Grolmes and H. K. Fauske, "Comparison of the Propagation Characteristics of Compression and Rarefaction Pressure Pulses in Two-Phase, One-Component Bubble Flow," *Trans. Am. Nucl. Soc.*, 11, 683 (1968).

85. V. V. Dvornichenko, *Teploenergetika*, 13, *10*, 72 (1966); see also, English translation, "The Speed of Sound in the Two-Phase Zone," *Thermal Eng.*, 13, *10*, 110 (1966).

86. R. E. Henry, M. A. Grolmes, and H. K. Fauske, "Propagation Velocity of Pressure Waves in Gas-Liquid Mixtures," Gas-Liquid Flow Symp., Waterloo University, Waterloo, Canada (1968).

87. R. E. Collingham and J. C. Firey, "Velocity of Sound Measurements in Wet Steam," *Ind. Eng. Chem. Process Des. Dev.*, 3, 197 (1963).

88. W. G. England, J. C. Firey, and O. E. Trapp, "Additional Velocity of Sound Measurements in Wet Steam," *Ind. Eng., Chem. Process Des. Dev.*, 5, 198 (1966).

89. M. E. Deich, G. A. Filippov, E. V. Stekol'Shchekov, and M. P. Anisimova, "Experimental Study of the Velocity of Sound in Wet Steam," *Thermal Eng.*, 14, *4*, 59 (1967).

90. V. J. DeJong and J. C. Firey, "Effect of Slip and Phase Change on Sound Velocity in Steam-Water Mixtures and Relation to Critical Flow," *Ind. Eng., Chem. Process Des. Dev.*, 7, 454 (1968).

91. R. E. Henry, "Pressure Wave Propagation in Two-Phase Mixtures," *Chem. Eng. Symp. Ser.*, 66, *102*, 1 (1970).

92. L. J. Hamilton and V. E. Schrock, "Propagation of Rarefaction Waves Through Two-Phase, Two-Component Media," *Trans. Am. Nucl. Soc.*, 11, 367 (1968).

93. A. R. Edwards, "Conduction Controlled Flashing of a Fluid and the Production of Critical Flow Rates in a One-Dimensional System," AHSB(S)R 147, U. K. Atomic Energy Authority, Risley, England (1968).

94. F. J. Moody, "Maximum Discharge Rate of Liquid-Vapor Mixtures from Vessels," in *Non-Equilibrium Two-Phase Flows*, R. T. Lahey, Jr. and G. B. Wallis, Eds., American Society of Mechanical Engineers, New York (1975).

95. G. L. Sozzi and W. A. Sutherland, "Critical Flow of Saturated and Subcooled Water at High Pressure," in *Non-Equilibrium Two-Phase Flows*, R. T. Lahey, Jr. and G. B. Wallis, Eds., American Society of Mechanical Engineers, New York (1975).

96. D. L. Caraher and T. L. DeYoung, "Interim Report on the Evaluation of Critical Flow Models," Aerojet Nuclear Company (1975).

97. H. K. Fauske, "Contribution to the Theory of Two-Phase, One-Component Critical Flow," ANL-6633, Argonne National Laboratory (1962).

98. S. M. Zivi, "Estimation of Steady-State Steam Void Fraction by Means of the Principle of Minimum Entropy Production," *Trans. ASME, Ser. C, J. Heat Transfer*, 86, 247 (1964).

99. J. E. Cruver and R. W. Moulton, "Critical Flow of Liquid-Vapor Mixture," *AIChE J.*, 13, 52 (1967).

100. H. K. Fauske, "What's New in Two-Phase Flow," *Power Reactor Technol.*, 9, 1, 35 (1966).

101. A. A. ARMAND, "The Resistance During the Movement of a Two-Phase System in Horizontal Pipes," AERE-Trans. 828, U. K. Atomic Energy Research Establishment, Harwell, England (1959).

102. F. J. Moody, "Maximum Flow Rate of Single-Component, Two-Phase Mixture," *Trans. ASME, Ser. C, J. Heat Transfer*, 87, 1, 134 (1965).

103. F. R. Zaloudek, "The Low Pressure Critical Discharge of Steam-Water Mixtures from Pipes," HW-68934, Rev., Hanford Works, Richland, Washington (1961).

104. J. G. Burnell, "Flow of Boiling Water Through Nozzles, Orifices, and Pipes," *Engineering*, 164, 572 (1947).

105. R. E. Henry and H. K. Fauske, "The Two-Phase Critical Flow of One-Component Mixtures in Nozzles, Orifices, and Short Tubes," *Trans. ASME, J. Heat Transfer, Ser. C*, 93, 179 (1971).

106. P. Lieberman and E. A. Brown, "Pressure Oscillations in a Water Cooled Nuclear Reactor Induced by Water-Hammer Waves," *Trans. ASME, Ser. D, J. Basic Eng.*, 82, 901 (1960).

107. V. L. Streeter and E. B. Wylie, *Hydraulic Transients*, McGraw-Hill Book Company, New York (1967).

108. D. R. Miller, "Critical Flow Velocities for Collapse of Reactor Parallel Plate Fuel Assemblies," *Trans. ASME, Ser. A*, 82, 83 (1960).

109. V. L. Streeter and E. B. Wylie, "Two and Three-Dimensional Fluid Transients," *Trans. ASME*, 90, 501 (1968).

110. S. Fabic, "Two- and Three-Dimensional Fluid Transients," *Trans. Am. Nucl. Soc.*, 14, 300 (1971).

111. "Quarterly Progress Report on Water Reactor Safety Programs Sponsored by the NRC Division of Reactor Safety Research," April-June 1976, TREE-NUREG-1004, Idaho National Engineering Laboratory (1976).

112. S. Fabic, "Computer Program WHAM for Calculation of Pressure, Velocity, and Force Transients in Liquid Piping Networks," 67-49R, Kaiser Engineers (1967).

113. W. D. Baines and E. G. Peterson, "An Investigation of Flow Through Screens," *Trans. ASME*, 73, 467 (1951).

114. H. C. Yeh and L. S. Tong, "Potential Flow Theory of a Pressurized Water Reactor," *Proc. Conf. Mathematical Models and Computational Techniques for Analyses of Nuclear Systems*, Ann Arbor, Michigan, CONF-730414-PI, U. S. Atomic Energy Commission, Washington, D. C. (1973).

115. H. C. Yeh, "Method of Solving the Potential Field in Complicated Geometries and the Potential Flow in the Lower Plenum of a Pressurized Water Reactor," *Nucl. Eng. Des.*, 32, 85 (1975).

116. G. Hetsroni, "Use of Hydraulic Models in Nuclear Reactor Design," *Nucl. Sci. Eng.*, 28, 1 (1967).

117. E. U. Khan, *Nucl. Technol.*, 16, 479 (1972).

118. R. J. Scavuzzo, "An Experimental Study of Hydraulically Induced Motion in Flat Plate Assemblies," WAPD-BT-25, p. 37, Bettis Atomic Power Laboratory (1962).

119. F. J. Remick, "Hydraulically Induced Deflection of Flat Parallel Fuel Plates," PhD Thesis, Pennsylvania State University (1963).

120. M. W. Wambsganss, "Second Order Effects as Related to Critical Coolant Flow Velocities and Reactor Parallel Plate Fuel Assemblies," *Nucl. Eng. Des.*, 5, 3, 268 (1967).

121. D. R. Miller and R. G. Kennison, "Theoretical Analysis of Flow Induced Vibration of a Blade Suspended in a Flow Channel," Paper No. 66-WAINE-1, ASME Annual Mtg., American Society of Mechanical Engineers, New York (Dec. 1966).

122. M. W. Wambsganss, "Flow Induced Vibration in Reactor Internals," *Power React. Technol. React. Fuel Process.*, 10, 1, 2 (1966-1967).

123. A. Roshko, "Experiments on Flow Past a Circular Cylinder at Very High Reynolds Number," *J. Fluid Mech.*, 10, 345 (1961).

124. H. Rouse, "Cavitation and Energy Dissipation in Conduit Expansion," *Proc. Int. Assoc. Hydraulic Research, 11th Congr.*, Leningrad, U.S.S.R., Vol. 1, Paper 1.28 (1965).

125. G. J. Bohm, "Natural Vibration of Reactor Internals," *Nucl. Sci. Eng.*, 22, 143 (1965).

126. G. Bowers and G. Horvay, "Beam Models of Vibration of Thin Cylindrical Shell Flexibly Supported and Immersed in Water Inside a Co-Axial Cylindrical Container of Slightly Larger Radius," *Nucl. Eng. Des.*, 26, 291 (1974).

127. L. E. Penzes, "Theory of Perry's Induced Pulsatory Coolant Pressure in Pressurized Water Reactors," *Nucl. Eng. Des.*, 27, 176 (1974).

128. G. Bowers and G. Horvay, "Forced Vibration of a Shell Inside a Narrow Annulus," *Nucl. Eng. Des.*, 34, 221 (1975).

129. N. K. Au-Yang, "Response of Reactor Internals to Fluctuating Pressure Forces," *Nucl. Eng. Des.*, 35, 361 (1975).

130. A. Powell, "Fatigue Failure of Structures Due to Vibrations Excited by Random Pressure Fields," *J. Acoust. Soc. Am.*, 30, 12 (1958).

131. D. Burgreen, J. J. Byrnes, and D. M. Benforado, "Vibration of Rods Induced by Water in Parallel Flow," *Trans. ASME*, 80, 991 (1958).

132. E. P. Quinn, "Vibration of Fuel Rods in Parallel Flow," GEAP-4059, General Electric Company (1962).

133. H. Sogreah, "Study of Vibrations and Load Losses in Tubular Clusters," Initial Special Report No. 3, EURATOM Report EURAEC-288, Société Grenobloise d'Etude et d'Applications Hydrauliques, Grenoble, France (1962).

134. R. T. Pavlica and R. C. Marshall, "Vibration of Fuel Assemblies in Parallel Flow," *Trans. Am. Nucl. Soc.*, 8, 599 (1965).

135. M. P. Paidoussis, "The Amplitude of Fluid-Induced Vibration of Cylinders in Axial Flow," AECL-2225, Atomic Energy of Canada Ltd. (Mar. 1965).

136. A. E. Morris, "A Review on Vortex Streets, Periodic Wakes, and Induced Vibration Phenomena," *Trans. ASME, J. Basic Eng.*, 85, *1*, 185 (1964).

137. M. P. Paidoussis, "An Experimental Study of the Vibration of Flexible Cylinders Induced by Axial Flow," *Trans. Am. Nucl. Soc.*, 11, 352 (1968).

138. W. C. Kinsel, "Flow Induced Vibrations of Spiral Wire Wrapped Assemblies," ASME Paper 75-WA/HT-76, American Society of Mechanical Engineers, New York (1975).

139. D. J. Gorman, "An Analytical and Experimental Investigation of the Vibration of Cylindrical Reactor Fuel Elements in Two-Phase Parallel Flow," *Nucl. Sci. Eng.*, 44, 277 (1971).

140. D. J. Gorman, "Experimental and Analytical Study of Liquid and Two-Phase Flow Induced Vibration in Reactor Fuel Bundles," ASME Paper 75-PVP-52, American Society of Mechanical Engineers, New York (1975).

141. M. Ruddick, "An Experimental Investigation of the Heat Transfer at High Rates Between a Tube and Water with Conditions at or near Boiling," PhD Thesis, University of London (1953).

142. W. H. Lowdermilk, C. D. Lanzo, and B. L. Siegel, "Investigation of Boiling Burnout and Flow Stability for Water Flowing in Tubes," Report TN-4382, National Advisory Committee for Aeronautics (1958).

143. F. Mayinger, O. Shad, and E. Weiss, "Research on the Critical Heat Flux (Burnout) in Boiling Water," Final Report, EURAEC-1811, EURATOM (1967).

144. G. B. Wallis, "Some Hydrodynamic Aspects of Two-Phase Flow and Boiling," *Proc. Int. Dev. Heat Transfer, Part II*, American Society of Mechanical Engineers, New York (1961).

145. D. H. Weiss, "Pressure Drop in Two-Phase Flow," AECU-2180, U. S. Atomic Energy Commission, Washington, D. C. (1952).

146. M. J. Saari, J. A. Heller, R. G. Dorsch, P. L. Stone, H. G. Hurrell, M. V. Gatstein, and C. H. Hauser, "Topics in Rankine Cycle Power Systems Technology," *Selected Technology for the Electric Power Industry*, NASA-SP-5057, pp. 35-90, National Aeronautics and Space Administration (1968).

147. M. Ledinegg, "Instability of Flow During Natural and Forced Circulation," AEC-tr-1861, U. S. Atomic Energy Commission translation from *Die Warme*, 61, 891 (1938).

148. H. H. Stenning and T. N. Veziroglu, "Flow Oscillation Modes in Forced Convection Boiling," *Proc. 1965 Heat Transfer and Fluid Mechanics Inst.*, Stanford University Press, Stanford, California (1965).

149. L. G. Neal and S. M. Zivi, "Hydrodynamic Stability of Natural Circulation Boiling System, Vol. 1: A Comparative Study of Analytical Models and Experimental Data," STL-372-14(1), TRW Systems, Redondo Beach, California (1965).

150. D. Moxin, "Stability of Once-Through Steam Generators," 73-WA/H7-24, American Society of Mechanical Engineers, New York (1973).

151. F. J. Edeskutz and R. S. Thurston, "Similarity of Flow Oscillation Induced by Heat Transfer in Cryogenic Systems," Symp. Two-Phase Flow Dynamics, Eindhoven, The Netherlands (1967).

152. A. J. Cornelius and J. D. Parker, "Heat Transfer Instabilities Near the Thermodynamic Critical Point," *Proc. Heat Transfer Fluid Mechanics Inst.*, Stanford University Press, Stanford, California (1965).

153. B. J. Walker and D. G. Harden, "Heat Driven Pressure and Flow Transients in the Supercritical Thermodynamics Region," Paper No. 64-WA/HT-37, American Society of Mechanical Engineers, New York (1964).

154. A. E. Bergles, P. Goldberg, and J. S. Maulbetsch, "Acoustic Oscillations in a High Pressure Single Channel Boiling System," Symp. Two-Phase Flow Dynamics, Eindhoven, The Netherlands (1967).

155. M. Ishii, *Thermo-Fluid Dynamic Theory of Two-Phase Flow*, Eyrolles, Paris, France (1975).

156. R. T. Lahey, Jr., "Two-Phase Flow in Boiling Water Nuclear Reactors," NEDO-13388, General Electric Company (1974).

157. S. L. Soo, *Fluid Dynamics of Multi-Phase Systems*, Blaisdell Publications (1967).

158. E. Quandt, "Analysis and Measurement of Flow Oscillations," *Chem. Eng. Prog., Symp. Ser.*, 57, *32*, 111 (1961).

159. A. B. Jones and A. G. Dight, "Hydrodynamic Stability of a Boiling Channel, Part II," KAPL-2208, Knolls Atomic Power Laboratory (1962).

160. C. L. Spigt, "On the Hydraulic Characteristics of a Boiling Water Channel with Natural Circulation," WW016-R92, Technological University of Eindhoven, The Netherlands (1966).

161. R. P. Anderson, L. T. Bryant, J. C. Carter, and J. F. Marchaterre, "Transient Analysis of Two-Phase Natural Circulation Systems," ANL-6653, Argonne National Laboratory (1962).

162. E. S. Beckjord and S. Levy, "Hydraulic Instability in a Natural Circulation Loop with Net Steam Generation at 1000 psia," GEAP-3215, General Electric Company (1959).

163. K. Becker, R. Mathisen, O. Eklind, and B. Norman, "Measurements of Hydrodynamic Instabilities, Flow Oscillations, and Burnout in a Natural Circulation Loop," S-316, EAES Symp. Two-Phase Flow, Studsvik, Sweden (1963).

164. K. C. Jain, "Self-Sustained Hydrodynamic Oscillations in a Natural Circulation Two-Phase Flow Boiling Loop," PhD Thesis, Northwestern University (1965).

165. J. Randles, "Kinetics of Boiling Hydraulic Loops," AEEW-R87, U. K. Atomic Energy Establishment, Winfrith, England (1961).

166. K. L. Garlid, "Theoretical Study of Transient Operation and Stability of Two-Phase Circulation Loop," PhD Thesis, University of Minnesota (1961).

167. A. N. Nahavandi and R. F. Von Hollen, "Flow Stability in Large Vertical Steam Generators," Paper 64-10 A/AUT-11, American Society of Mechanical Engineers, New York (Dec. 1964).

168. M. Ishii, "Thermally Induced Flow Instabilities in Two-Phase Mixtures in Thermal Equilibrium," PhD Thesis, School of Mechanical Engineering, Georgia Institute of Technology (1971).

169. P. Saha, M. Ishii, and N. Zuber, "An Experimental Investigation of Thermally Induced Flow Oscillations in Two-Phase Systems," Paper 75-WA/HT-6, American Society of Mechanical Engineers, New York (1975).

170. H. B. Currin, C. M. Hunin, L. Rivlin, and L. S. Tong, "HYDNA-Digital Computer Program for Hydrodynamic Transients in a Pressure Tube Reactor or a Closed Channel Core," CVNA-77, U. S. Atomic Energy Commission, Washington, D. C. (1961).

171. N. Mitsuishi, S. Sakata, Y. Matsuda, Y. Yamamoto, and Y. Oyama, "Studies on Liquid Entrainment," AEC-tr-4225, U. S. Atomic Energy Commission, Washington, D. C. (1961).

172. D. M. Newitt, N. Dombrowski, and F. H. Knelman, "Liquid Entrainment 1. The Mechanism of Drop Formation from Gas or Vapor Bubbles," *Proc. Trans. Inst. Chem. Eng.*, **32**, 244 (1954).

173. L. Sterman, "Theory of Steam Separation," *I. Tech. Physics (USSR)*, **28**, 7 (1958).

174. R. F. Davis, "The Physical Aspect of Steam Generation at High Pressures and the Problem of Steam Contamination," *Proc. I. Mech. Eng.*, **144**, 1198 (1940).

175. J. Wilson and M. McDermott, "Moisture-De-Entrainment Tests in Two and Four Inch Diameter Test Sections," ACNP-5921, Allis Chalmers Manufacturing Company (1959).

176. B. E. Bingham and K. C. Shieh, "REFLOOD—Description of Model for Multinode Reflood Analysis," BAW-10093, Babcock & Wilcox Company (1974).

177. D. J. Hanson, C. E. Cartmill, K. R. Perkins, C. J. Shaffer, and D. M. Snider, "ECC Performance in the Semi-Scale Geometry," ANCR-1161, Aerojet Nuclear Company (1973).

178. R. A. Cudnick and R. O. Wooton, "Performance of Injected ECC Water Through the Downcomer Annular in the Presence of Reversed Core Steam Flow," Battelle Columbus Laboratories (1974).

179. C. J. Crowley, G. B. Wallis, and D. L. Ludwig, "Steam-Water Interaction in a Sealed Pressurized Water Reactor Downcomer Annular," COO-2294-4, Dartmouth College (1974).

180. J. A. Block, G. B. Wallis, and C. J. Crowley, "Analyses of ECC Delivery," CREARE-TN-231, Creare Corporation, Dartmouth, New Hampshire (1976).

181. J. A. Block and C. J. Crowley, "Effect of Steam Upflow and Superheated Water on ECC Delivery in a Simulated Multiloop PWR Geometry," TN-210, Creare Corporation, Dartmouth, New Hampshire (1975).

182. W. L. Haberman and R. K. Morton, "An Experimental Investigation of the Drag and Shape of Air Bubbles Rising in Various Liquids," DTMB-802, David W. Taylor Model Basin, Hoboken, New Jersey (1953).

4

HEAT TRANSFER AND TRANSPORT

The reactor coolant removes heat from the core by transferring heat from the fuel elements through convection or boiling, and then transporting the energy to the steam generator. Since the rate of heat transfer depends on the enthalpy of the fluid as well as on the heat flux, we first examine the determination of fluid enthalpy. Expressions useful in describing the axial heat-flux distribution in a reactor core are developed and then used to compute the amount of heat transported by the coolant as a function of distance through the core. The heat transfer mechanisms that apply during single- and two-phase flow are presented, and solid moderator and core structure cooling is considered. Extensive tables of water and steam properties are available elsewhere[1,2] and are not repeated here.

4-1 HEAT TRANSPORT

4-1.1 Steady-State Enthalpy Rise Along a Closed Coolant Channel

In a pressure tube reactor, there is no fluid interchange with surrounding channels, and the average enthalpy rise for the entire tube can be obtained by a simple heat balance. In a vessel-type reactor, mixing and cross flow affect the behavior of a single subchannel. However, when the average enthalpy rise along an entire assembly is sought, these effects are of lesser importance. Thus, a rough approximation of the average enthalpy rise along a coolant channel is often obtained considering the channel to be closed. Such an approach is taken in some of the computer codes that reactor operators use to evaluate steady-state core performance.[3]

The heat absorbed by the coolant at location Z can be evaluated by taking a heat balance around a differential element and then integrating along the core. When the interchange with the surrounding region can be ignored, we get

$$(A/p')G(\Delta H) = \int_{-L/2}^{Z} q''(z)dz \ , \tag{4.1}$$

where

$q''(z)$ = assembly-average heat flux at axial location z, Btu/h ft^2

A = flow area of assembly, ft^2

p' = total heated perimeter of assembly, ft

ΔH = enthalpy at location z less inlet enthalpy, Btu/lb

G = mass flow rate, lb/h ft^2

L = core length.

In a large assembly, $(A/p') = (D_e/4)$.

The heat flux at a given axial location can be expressed in terms of the hot-channel factor, $F_{xy}(z)$. We then have

$$(A/p')G(\Delta H) = \int_{-L/2}^{Z} F_{xy}(z)q''_{av}\,dz \ , \tag{4.2}$$

where q''_{av} is the average heat flux in the core. In many cases, no analytical expressions are available for $F_{xy}(z)$, and enthalpy rise is approximated from

$$(A/p')G(\Delta H) = q''_{av} \sum_{j} [F_{xy}(z)]_j (\Delta z)_j \ , \tag{4.3}$$

where

$F_{xy}(z)_j$ = average value of $F_{xy}(z)$ in axial segment j

$(\Delta z)_j$ = length of axial segment j.

4-1.1.1 Behavior with Simple Axial Flux Shapes

The axial power distribution of an unperturbed cylindrical reactor core was shown in Chap. 1 to follow a cosine curve; therefore, local flux q'' can be expressed as a function of q''_{max}, the peak flux.

As shown in Fig. 4-1, we define

L_0 = length between $\theta = 0$ and $\theta = \pi$ (half of cosine cycle length)

$L = L_0 + L'$ = actual core length

L' = difference between actual core length and half of cosine cycle length (negative for a chopped cosine)

z = axial distance from center line.

When $F_Z^N \leqslant 1.57$, we have

$$q'' = q''_{max} \cos \theta = q''_{max} \cos(\pi z/L_0) \ . \tag{4.4}$$

If we replace $q''(z)$ in Eq. (4.1) by its equivalent from Eq. (4.4), for the enthalpy rise we obtain

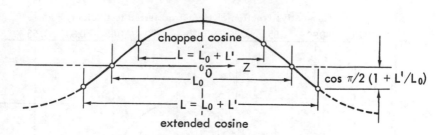

Fig. 4-1.   Approximate axial-flux distribution within thermal reactor cores.

$$\Delta H = (p/AG) \int_{-L/2}^{Z} q''_{max} \cos(\pi z/L_0) dz \quad . \tag{4.5}$$

If the cumulative fraction F of heat absorbed by the coolant is defined as the ratio of heat absorbed at location Z to the total heat generated in the channel, then F can be evaluated by

$$F = \frac{\int_{-L/2}^{Z} q''_{max} \cos(\pi z/L_0) dz}{\int_{-L/2}^{+L/2} q''_{max} \cos(\pi z/L_0) dz} \quad , \tag{4.6}$$

providing the axial distribution is cosinusoidal. With this definition of F, the enthalpy of fluid H at location Z is given by

$$H = H_{in} + [4 \, q''_{max} \, L/(F_Z^N D_e G)]F \quad , \tag{4.7}$$

where H_{in} is inlet enthalpy. Values of F at various distances along a reactor channel are plotted in Fig. 4-2 for several values of L/L_0.

Under most circumstances, values of L_0 are not given, but values of F_Z^N are stated. However, by using our previous definition of axial hot-channel factor F_Z^N, and referring to Fig. 4-2, we express F_Z^N in terms of L' and L_0. For a chopped cosine distribution, we obtain

$$F_Z^N = \frac{1}{\dfrac{2}{L} \displaystyle\int_{0}^{L/2} \cos \dfrac{\pi z}{L_0} dz} = \frac{\pi \left(1 + \dfrac{L'}{L_0}\right)}{2 \sin\left[\dfrac{\pi}{2}\left(1 + \dfrac{L'}{L_0}\right)\right]} \quad , \tag{4.8}$$

where $-L/2 \leqq z \leqq L/2$ and L' is negative.

For cosine heat flux distribution with reported axial hot-channel factors in excess of 1.57 $(\pi/2)$, L' becomes positive. The flux is assigned a value of zero at $Z = \pm L/2$. We then have

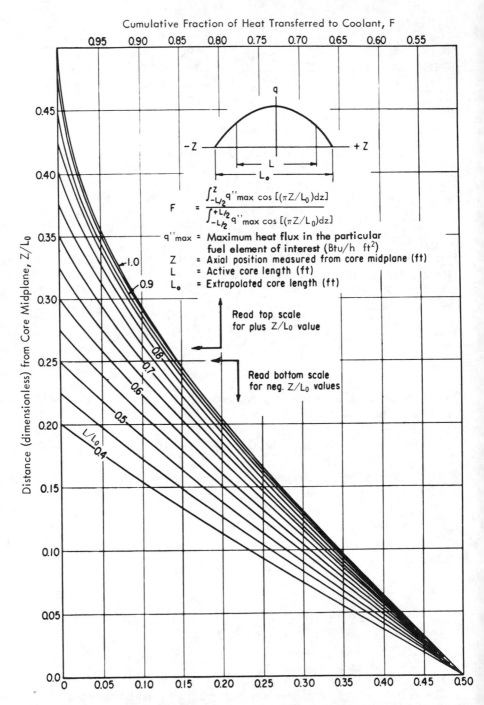

Fig. 4-2.   Heat absorbed by coolant along reactor channel. [From *Nucleonics*, **18**, *11*, 170 (1960).]

$$q'' = q''_{max} \frac{\cos\dfrac{\pi z}{L_0} - \cos\left[\dfrac{\pi}{2}\left(1 + \dfrac{L'}{L_0}\right)\right]}{1 - \cos\left[\dfrac{\pi}{2}\left(1 + \dfrac{L'}{L_0}\right)\right]} \, , \tag{4.9}$$

where $-L/2 \leqq z \leqq L/2$. Hence, for an extended cosine distribution,

$$F_Z^N = \frac{1 - \cos\left[\dfrac{\pi}{2}\left(1 + \dfrac{L'}{L_0}\right)\right]}{\dfrac{2}{L}\displaystyle\int_0^{L/2}\left\{\cos\dfrac{\pi z}{L_0} - \cos\left[\dfrac{\pi}{2}\left(1 + \dfrac{L'}{L_0}\right)\right]\right\}dz}$$

$$= \frac{1 - \cos\dfrac{\pi}{2}\left(1 + \dfrac{L'}{L_0}\right)}{\dfrac{2}{\pi\left(1 + \dfrac{L'}{L_0}\right)}\sin\left[\dfrac{\pi}{2}\left(1 + \dfrac{L'}{L_0}\right)\right] - \cos\left[\dfrac{\pi}{2}\left(1 + \dfrac{L'}{L_0}\right)\right]} \, . \tag{4.10}$$

The values of L'/L_0 corresponding to the assigned value of F_Z^N are plotted in Fig. 4-3. Note that since two separate equations are used for defining F_Z^N, there is a change in slope at $L'/L_0 = 0$.

As noted in Sec. 1-2.3, a better approximation of an axial flux distribution that has been skewed by rod insertion is sometimes given by

$$q'' = [A + (Bz/L)]\cos(\alpha_0 z/L) \, . \tag{4.11}$$

The heat flux is assumed to go to zero at $z/L = \pi/2\alpha_0$. By assuming that the slope of the flux curve is also zero at this value of z/L, constants A and B can be obtained in terms of q''_{max}, and we have

$$q'' = 0.54954 \, q''_{max}\left(\frac{\pi}{2} - \frac{\alpha_0 z}{L}\right)\cos\left(\frac{\alpha_0 z}{L}\right) \, , \tag{4.12}$$

and

$$F_Z^N = (1.1585) \, \alpha_0/\sin \alpha_0 \, . \tag{4.13}$$

Equation (4.13) can be used to determine α_0 from the value of F_Z^N. If a distribution skewed toward the top of the core is being considered, the above relations apply with each value of z/L replaced by its negative value. The foregoing applies as long as F_Z^N is <1.82. At higher values, negative fluxes at the core ends will be encountered.

4-1.2 Steady-State Enthalpy Rise Along a Semi-Open Channel

While the assumption of essentially zero interchange with adjacent channels can, under some circumstances, be considered approximately correct when the

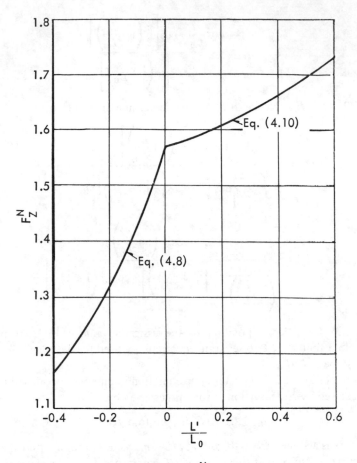

Fig. 4-3.   Plot of F_Z^N versus L'/L_0.

average behavior of a fuel assembly is determined, the assumption is not appropriate when we consider the channels defined by four neighboring rods in a fuel assembly. Here interchannel mixing and cross flow should always be considered.

Adjacent subchannels are open to each other through the gap between two neighboring fuel rods; flow in one channel mixes with that of the other. In addition, as previously observed, there is also cross flow between channels because of the pressure gradient. Local turbulent mixing reduces the enthalpy rise of the hot channel. On the other hand, flow leaving the hot channel increases its enthalpy rise. Calculation of the net result is complicated although the equation describing enthalpy rise is easily written.

For simplicity, for a segment of length ΔZ we consider the case of two adjacent subchannels that are only linked with each other. This situation is represented in Fig. 4-4. Quantities H, V, ρ, and P represent coolant enthalpy, velocity, density,

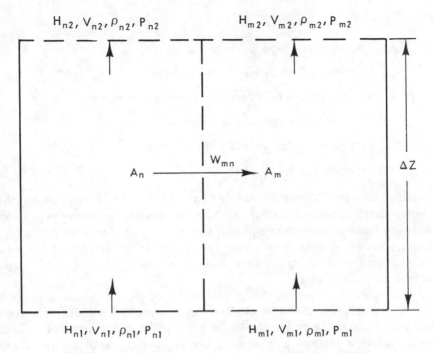

Fig. 4-4.   Mathematical representation of flow redistribution.

and static pressure; A is channel flow area; W_{mn} is cross flow between channels; and w' is the flow exchange of diffusion mixing. The numbered subscripts refer to two different axial elevations in the core. Writing mass, energy, and momentum balance equations for channel m between elevations 1 and 2 results in the following relationships where W_{mn} has been considered positive in the direction from channel n into channel m:

Conservation of Mass

$$A_m V_{m1} \rho_{m1} + W_{mn} = A_m V_{m2} \rho_{m2} \ . \tag{4.14}$$

Conservation of Heat

$$A_m V_{m1} \rho_{m1} H_{m1} + Q_{mz} + W_m \overline{H}_n + w'(H_n - H_m)\Delta Z = A_m V_{m2} \rho_{m2} H_{m2} \ . \tag{4.15}$$

Conservation of Axial Momentum

$$A_m P_{m1} + A_m \rho_{m1} V_{m1}^2 / g_c + W_{mn} \overline{V}_n / g_c$$
$$= A_m P_{m2} + A_m \rho_{m2} V_{m2}^2 / g_c + K_{mz} A_m \overline{\rho}_m V_m^2 / 2 g_c + \overline{\rho}_m \Delta Z g / g_c \ , \tag{4.16}$$

where

Q_{mz} = heat input into channel m in interval ΔZ

W_{mn} = cross flow from channel n to channel m

$\overline{H}, \overline{\rho}, \overline{V}$ = mean values in ΔZ of enthalpy, density, and velocity

w' = flow exchange rate per unit length

K_{mz} = pressure loss coefficient for channel m in interval ΔZ

P = pressure.

Evaluation of ρ and K_{mz} requires the determination of void fraction and two-phase pressure drop as described in Chap. 3. Cross flow is determined from the appropriate lateral momentum balance equation (see Sec. 3-1.3). The interchange due to mixing, represented by w', is determined by a suitable correlation (see Sec. 3-1.3). The additional terms arising from the coupling of channel m to other surrounding channels are readily deduced.

A similar set of equations can be written for every other channel in the region studied. A simultaneous solution of all of these equations is required to determine fluid conditions at the exit of the length interval. The complexity of the calculational procedure requires a computer solution; a number of computer codes have been written for this purpose. In all of these codes, the cross section of the region of interest is divided into a series of subchannels, while the length is divided into a series of axial segments. Thus, a set of connecting control volumes is provided. All of the calculational procedures assume that conditions are uniform within a given control volume. Gradients exist only across control volume boundaries.

Earlier computer codes, such as THINC II (Refs. 4 and 5), COBRA II (Ref. 6), and HAMBO (Ref. 7) use "marching procedures." If conditions at the inlet of a given axial segment are known, the governing equations can be solved iteratively to give the conditions at the exit of the segment. By proceeding stepwise along the channel length, a marching solution for the whole channel is obtained.

The iterative-interval calculation methods in COBRA-II and HAMBO are similar. A set of cross flows between subchannels is "guessed." With this set, the energy equation is solved by forward differencing in COBRA-II and central differencing in HAMBO. From the momentum equation, pressure drop in each subchannel is calculated. From this calculation, a new set of cross-flow guesses that give a pressure balance is obtained by backward differencing. The iteration is continued until an acceptable pressure balance is obtained.

The boundary conditions to be satisfied are that the lateral pressure difference between subchannels should be zero at channel inlet and exit. Having passed once along the channel, this implies that iteration over the channel length may be necessary by using improved guesses of flow division between subchannels at the inlet. In practice, only one pass may be necessary, particularly for a hydraulic model in which lateral momentum transfer is neglected or only notionally included.

Rowe[8] has shown that for a single-pass solution to be stable and acceptable, the calculational increment of length must be greater than a critical value

$$\frac{2C\,|w|}{g(m,H)} \, ,$$

where

C = cross-flow resistance coefficient

w = cross flow

$g(m,H)$ = difference in axial pressure gradient caused by the cross flow

m = subchannel flow rate

H = enthalpy.

For long enough increments, calculated exit conditions for an interval tend to compensate for errors in the assumed inlet conditions. This provides a self-correcting mechanism in the calculation and, conversely, means that large changes in assumed channel inlet conditions are required to affect calculated conditions and pressure balance at the channel exit.

The acceptability of a single-pass marching solution depends on coupling between subchannels. If this is weak (e.g., if the cross flows are small), a single-pass marching solution technique is adequate; this follows from Rowe's criterion above. Upstream effects would be entirely confined to the preceding interval. From Rowe's criterion, stronger coupling (i.e., large cross flows) could mean the intervals need to be impracticably large.

A multipass marching solution is used in COBRA IIIC (Ref. 9). The inlet flow division between subchannels is fixed as a boundary condition, and an iterated solution is obtained to satisfy the other boundary solution of zero pressure differential at channel exit. The procedure is to guess a pattern of subchannel boundary pressure differentials for all mesh points simultaneously and from this to compute, without further iteration, the corresponding pattern of cross flows using a marching technique up the channel. The pressure differentials are updated during each pass, and the overall channel iteration is completed when the fractional change in subchannel flows is less than a preset amount.

Pressure differentials at the exit of the length steps are used to calculate cross flows. Cross flows are then used to calculate pressure differentials at the outlet of the previous (upstream) length steps. These pressure differentials are saved for use during the next iteration. At the exit of the last length step, the boundary condition of zero pressure differential is imposed and cross flows at the exit are calculated on this basis. The iterative procedure forces agreement with the assumed boundary condition.

The TORC code[10] is a modified version of COBRA IIIC. The basic numerics are those of COBRA IIIC, but TORC contains some additional features useful in overall core design.

The THINC II code[4,5] handles the problem of zero lateral pressure gradient at the assembly exit by assuming that the lateral pressure gradient is zero everywhere. Chelemer et al.[5] argue that because of close coupling between subchannels, there can be only a very low pressure gradient in an assembly. Under these conditions, they show that changes in the lateral pressure drop cause very small differences in axial flows. Hence, the change in axial flow in a given channel is determined by the requirement that pressure drop be the same across each control volume at a given elevation. Since pressure gradients are not given, enthalpy and axial velocity associated with cross flow are not directly calculable. Therefore, these are taken as a weighted average of the values of the surrounding channels. Weighting factors depend on control volume net gain or loss in flow over the length step.

Use of a marching solution to determine the behavior of individual subchannels in an assembly requires that the inlet flow to that assembly be known. The assumption that all assemblies in a core have the same inlet flow can be appreciably in error. Flow must be divided so that the core pressure drop remains essentially constant. Therefore, higher pressure loss coefficients in high power assemblies, due to the presence of significant exit quality, will lead to lower flows in these assemblies.

The THINC I code[11] was the first calculational technique capable of satisfactorily assigning inlet flows to the assemblies within a semi-open core. In the THINC I approach, we recognize that the total pressure distribution at the top of the core region is a function of inlet pressure, density, and velocity distributions. This functional dependence can then be expressed as

$$P_j^o = P_j^o(P_1^i, P_2^i, \ldots, P_n^i, \rho_1^i, \rho_2^i, \ldots, \rho_n^i, V_1^i, V_2^i, V_n^i)$$

$$j = 1, 2, \ldots, n \ , \tag{4.17}$$

where superscripts i and o represent inlet and outlet values, respectively, and j is the number of the channel.

Since the flow leaving the core enters a large plenum across which significant pressure differentials do not exist, the basic criteria to be satisfied are that outlet total pressure is uniform and total mass flow rate must be constant; i.e.,

$$P_1^o = P_2^o = P_3^o = \ldots = P_n^o \ , \tag{4.18}$$

and

$$\sum_{j=1}^{n} A_j \rho_j^i V_j^i = \text{a constant} \ . \tag{4.19}$$

As flow enters the core from a large plenum with only small pressure differences across it, these basic criteria can only be achieved by adjusting the inlet velocity distribution, V_j^i.

Let us consider a two-channel core, for example: If inlet pressures and densities are fixed,

$$P_1^o = \phi(V_1, V_2)$$
$$P_2^o = \psi(V_1, V_2) \ ,$$
(4.20)

then

$$dP_1^o = \phi_1 dV_1 + \phi_2 dV_2$$
$$dP_2^o = \psi_1 dV_1 + \psi_2 dV_2 \ ,$$
(4.21)

where

$$\phi_1 = \frac{\partial \phi}{\partial V_1} \ , \qquad \phi_2 = \frac{\partial \phi}{\partial V_2}$$
$$\psi_1 = \frac{\partial \psi}{\partial V_1} \ , \qquad \psi_2 = \frac{\partial \psi}{\partial V_2} \ ,$$
(4.22)

and

$$[dP^o] = \begin{bmatrix} \phi_1 \ \phi_2 \\ \psi_1 \ \psi_2 \end{bmatrix} [dV^i] \ .$$
(4.23)

Let

$$[T] = \begin{bmatrix} \phi_1 \ \phi_2 \\ \psi_1 \ \psi_2 \end{bmatrix}$$
(4.24)

and

$$[T^{-1}] = \begin{bmatrix} \dfrac{\psi_2}{\phi_1 \psi_2 - \phi_2 \psi_1} & \dfrac{-\phi_2}{\phi_1 \psi_2 - \phi_2 \psi_1} \\ \dfrac{-\psi_1}{\phi_1 \psi_2 - \phi_2 \psi_1} & \dfrac{\phi_1}{\phi_1 \psi_2 - \phi_2 \psi_1} \end{bmatrix} \ .$$
(4.25)

Thus, for a given inlet pressure distribution and inlet density distribution, the following applies:

$$[dP^o] = [T] [dV^i] \ ,$$
(4.26)

where

$[dP^o]$ = column matrix with elements dP_j^o

$[T]$ = n by n matrix with elements $\partial P_j^o / \partial V_k^i$

$[dV^i]$ = column matrix with elements dV_k^i.

For square matrices, since $[T] [T^{-1}] = I$, an identity matrix, premultiplication of Eq. (4.26) by $[T^{-1}]$ gives

$$[dV^i] = [T^{-1}] [dP^o]$$
(4.27)

provided $[T]$ is not singular.

Consider the pressure residues at the region outlet to be given by the relation

$$\Delta P_j^o = \bar{P}^o - P_j^o \ , \tag{4.28}$$

where $\bar{P}^o$ is the average outlet total pressure. Let the changes in velocities ΔV_k^i associated with pressure residues ΔP_j^o be given by a relation analogous to Eq. (4.27)

$$[\Delta V^i] = [T^{-1}]\{[\Delta P^o] - \lambda[1]\} \ , \tag{4.29}$$

where

$[\Delta V^i]$ = column matrix with elements ΔV_k^i

$[\Delta P^o]$ = column matrix with elements ΔP_j^o

$[1]$ = unit column matrix

λ = scalar quantity to be determined.

Equation (4.19) allows us to determine λ. If the initial choice of velocities satisfies Eq. (4.19) and all subsequent choices satisfy this constraint, then

$$[A][\rho][\Delta V^i] = 0 \ , \tag{4.30}$$

where $[A]$ is a row matrix with elements A_j, $[\rho]$ is a square matrix with leading diagonal ρ_j^i, and all other elements are zero. Combining Eqs. (4.29) and (4.30) and solving for λ,

$$\lambda = [A][\rho][T^{-1}][\Delta P^o]/\{[A][\rho][T^{-1}][1]\} \ . \tag{4.31}$$

Thus, if the $[T]$ matrix is known and is nonsingular, λ can be calculated from Eq. (4.31) and substituted in Eq. (4.29) to give the changes in inlet velocities associated with outlet pressure distribution P_j^o.

To obtain $[T]$, the inlet velocity to one channel is changed by a preassigned small fraction, ϵ, and the resulting changes in outlet pressures for all the channels are determined by the step-wise procedure described above. The elements of the first column of the $[T]$ matrix are determined from this calculation. The original velocity distribution is restored, and the procedure is repeated for each channel. Thus, the inlet velocity to each channel is changed by the same small fraction, ϵ, and in turn, the elements of each column of the $[T]$ matrix are determined. The inverse of the $[T]$ matrix is then determined and λ is calculated using Eq. (4.31). Substituting in Eq. (4.29) gives matrix $[\Delta V^i]$, which is then used to determine a new inlet velocity distribution using the relation

$$[V'^i] = [V^i] + [\Delta V^i] \ , \tag{4.32}$$

where $[V^i]$ and $[V'^i]$ are the column matrices for the original and new inlet velocities, respectively. This entire procedure is repeated until the error in the outlet pressure distribution is less than a preassigned value.

Sha et al.[12] observe that under some conditions the foregoing procedure may not converge. Experience shows that the procedure always converges when there is no cross flow between channels. Advantage can be taken of this fact and a solution can be obtained for an infinite lateral resistance (or zero lateral flow area) in the core. The velocity distribution so obtained is then used as input to a second problem with a finite lateral flow resistance (or finite lateral flow area). This velocity distribution is then used to solve a further problem in which the lateral resistance or flow area more closely approaches the actual value. The process continues until convergence is obtained at the actual lateral resistance or flow area.

The convergence difficulty encountered when a marching solution is used to examine full core behavior is one of the motivating factors that led to the development of other solution procedures. An additional motivation was the desire to be able to handle recirculating flows that could develop under severe blockage conditions. Marching solutions cannot handle reverse flows.

Reverse flows can be successfully treated by numerical procedures that solve the conservation equations for the control volumes at all axial levels simultaneously. This approach is followed in THINC IV code.[13,14] Here, lateral velocity components are regarded as perturbed quantities much smaller than axial flow velocity. The original governing equations are split into a perturbed and unperturbed system of equations. Perturbed momentum and continuity equations are then combined to form a field equation that is solved for the entire velocity field. Inlet velocities are determined such that a uniform outlet pressure is obtained. The initial solution obtained is used to update the properties and conditions assumed to exist in various control volumes. The iteration continues until assumed and calculated properties are in satisfactory agreement. The THINC IV code[13,14] is written so it can also be used for determining behavior in a core region where the subchannels are defined by four neighboring fuel rods.

The COBRA IV code[15] is another program that can treat the core-wide problem successfully. As in COBRA IIIC (Ref. 9), a multiple-pass marching solution is used. However, inlet flows are not set as a boundary condition, and outlet pressure differentials are not set at zero when exit cross flows are computed. Instead, the requirement that outlet pressure differentials be zero is met by adjusting the inlet flows. That is, if flow $(w_i)_0$ to channel i leads to pressure drop ΔP_i across channel i, flow to channel i in the next iteration, $(w_i)_1$, is given by

$$(w_i)_1 = \left\{ 1 - \frac{(\Delta P_s - \Delta P_i)}{2(\Delta P_i - \bar{h})} \right\} (w_i)_0 \ , \qquad (4.33)$$

where

P_s = specified pressure drop across core

$\bar{h}$ = average gravitational head of all channels.

Castellana and Bonilla[16] compared rod bundle pressure drop data with pressure drop calculations obtained from subchannel modeling codes. They conclude that estimates from subchannel codes are superior to those from mixed flow models, particularly in tightly spaced bundles at low exit qualities. Weisman and Bowring[17] provide a comprehensive review of modeling procedures and computational techniques.

4-1.2.1 Advanced Computational Techniques

The development of more sophisticated analytical techniques is leading to further improvements in subchannel analysis computer programs. For example, Patankar and Spalding[18] have developed the SIMPLE algorithm that is now incorporated in the SABRE code.[19] In this algorithm, the pressure field is determined by first calculating an intermediate velocity field based on the estimated pressure field. An appropriate correction is then obtained to satisfy the continuity equation. The corrections are made on a line-by-line basis rather than point-to-point, and the solution is considerably speeded when there are a large number of nodes being considered.

Stewart and Rowe[20] extended the ICE technique developed by Harlow and Amsden.[21] Stewart and Rowe[20] use an iterative procedure for solving the conservation equations in which error in the energy equation is used to give appropriately updated values of density, mass flow, and pressure. The scheme was successfully incorporated in a version of COBRA IV used to study a PWR core during reflooding oscillations. The approach is said to avoid the computational difficulties that arise for liquid-vapor mixtures with large density ratios. Further computational advances can be expected to lead to increased accuracy and flexibility in the types of problems handled, as well as decreased computation time.

4-1.3 Transient Heat Transport

For a single closed channel, the partial differential equation describing the change in coolant enthalpy H with position Z and time t is again readily obtained by an energy balance around a differential control volume. For constant ρ, this yields

$$\rho \, \frac{\partial U}{\partial t} + G \, \frac{\partial H}{\partial z} = \frac{q'' p'}{A} \,, \tag{4.34}$$

when

q'' = heat flux, Btu/(h ft^2)

A = channel flow area, ft^2

p' = heated perimeter of the channel, ft

ρ = fluid density, lb/ft^3

U = internal energy at position Z, Btu/(lb ft^3).

When the system pressure remains constant, Eq. (4.34) can be simplified to

$$\rho \, \frac{\partial H}{\partial t} + G \, \frac{\partial H}{\partial z} = \frac{q'' p'}{A} \,. \tag{4.35}$$

For most situations, ρ varies significantly with H, and an analytic solution is not feasible. The usual procedure is to express Eq. (4.34) or (4.35) in finite difference form to obtain a numerical solution. When the coolant remains substantially subcooled, density changes can be negligible and an analytical solution for flow transients is then possible using the method of "characteristics."

4-1.3.1. Simple Flow and Power Transients

The method of characteristics is applicable to first-order linear partial differential equations of the form

$$S \frac{\partial z}{\partial x} + Q \frac{\partial z}{\partial y} = R \ , \tag{4.36}$$

where S, Q, and R are functions of x and y. It can be shown that the general solution of Eq. (4.36) is of the form

$$F(U_1, U_2) = 0 \ , \tag{4.37}$$

where $U_1(x,y,z) = C_1$ and $U_2(x,y,z) = C_2$ are any two independent solutions of

$$\frac{dx}{S} = \frac{dy}{Q} = \frac{dz}{R} \ . \tag{4.38}$$

The intersection of U_1 and U_2 is a characteristic curve whose tangent has the direction numbers $S{:}Q{:}R$.

We consider a flow transient where mass velocity G varies in accordance with

$$G = G_0/(1 + t) \ , \quad t > 0 \ , \tag{4.39}$$

where G_0 is a constant, and both heat flux and inlet enthalpy remain constant with time t. The basic differential equation is given by Eq. (4.35). Therefore, our auxiliary equation, which is analogous to Eq. (4.38), is

$$\frac{dt}{\rho} = \frac{dz}{G} = dH/(q''p'/A) \ . \tag{4.40}$$

There are two solutions to this differential equation. We first consider the behavior of a packet of fluid within the reactor when the transient began. It can be described in terms of position Z_0, which it had when the transient began. The subsequent positions such a given packet (e.g., Z_{0_1}) will hold are shown by the appropriate lines of Region I in Fig. 4-5. A packet of fluid that had not yet entered the reactor when the transient began can be described in terms of time t_0, measured from the beginning of the transient when the packet enters the reactor. The position versus time history of a designated packet is shown by a line within Region II of Fig. 4-5. The equations describing the solutions for Regions I and II can be expected to differ. The two regions are separated by the limiting characteristic corresponding to the history of the fluid packet that was just at the reactor inlet when the transient began. We find the solution for Region I by first integrating

$$dt/\rho = dz/G \tag{4.41}$$

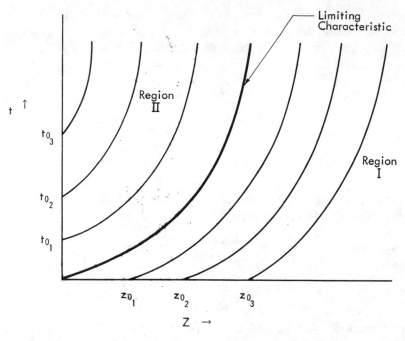

Fig. 4-5.   Time-distance relationships during a hypothetical flow transient.

to obtain

$$z - Z_0 = \int_0^t \frac{G_0}{\rho(1 + t)} dt = \frac{G_0}{\rho} \ln(1 + t) \ ,$$ (4.42)

where Z_0 is the axial position of the fluid packet at time zero. From integrating $dz/G = dH/(q''p/A)$, we obtain

$$H - H_0(Z_0) = \int_{Z_0}^{Z} \frac{(1 + t)}{G_0} \left(\frac{q''p'}{A} \right) dz \ ,$$ (4.43)

where $H_0(Z_0)$ is coolant enthalpy at zero time and axial position Z_0. We evaluate this quantity from steady-state conditions. By substituting into Eq. (4.43), we obtain

$$H_{(Z)} = H_{\text{in}} + \frac{p'}{AG_0} \int_0^{Z_0} q'' dz + \frac{p'}{AG_0} \int_{Z_0}^{Z} (1 + t) q'' dz \ ,$$ (4.44)

where H_{in} is inlet enthalpy. From Eq. (4.42), we have

$$(1 + t) = \exp[\rho(Z - Z_0)/G_0] \ .$$ (4.45)

Therefore,

$$H_{(Z)} = H_{in} + \frac{p'}{AG_0} \int_0^{Z_0} q'' dz + \frac{p'}{AG_0} \int_{Z_0}^{Z} q'' \exp[\rho(Z - Z_0)/G_0] dz . \quad (4.46)$$

Integrating the equation for the desired flux shape and substituting the value of Z obtained from Eq. (4.42) provides the desired solution for Region I.

To obtain the solution for Region II, we again begin by integrating Eq. (4.41), but now we consider the fluid that first entered the reactor after the transient began. Since t_0 is the time at which a given packet enters the reactor, we have

$$z = (G_0/\rho) \int_{t_0}^{t} dt/(1 + t) = (G/\rho) \ln[(1 + t)/(1 + t_0)] , \quad (4.47)$$

and by rearrangement,

$$(1 + t) = (1 + t_0)\exp(\rho z/G_0) . \quad (4.48)$$

Now, from

$$dz/G = dH/(q''p'/A) \quad (4.49)$$

$$H_{(Z)} - H_{in} = [p'/(AG_0)] \int_0^{Z} q''(1 + t)dz , \quad (4.50)$$

and by substituting for $(1 + t)$, we obtain

$$H_{(Z)} = H_{in} + \frac{p'}{G_0 A}(1 + t_0) \int_0^{Z} \exp(\rho z/G_0)q'' dz . \quad (4.51)$$

The limiting characteristic is obtained by replacing G in terms of G_0 and t in Eq. (4.41) and integrating the resulting equation for both z and t:

$$\int_0^{T} \frac{dt}{1 + t} = \int_0^{Z} \frac{\rho}{G_0} dz$$

$$\ln(1 + t) = \frac{\rho Z}{G_0} . \quad (4.52)$$

For $t \leqslant T$, the Region I solution, Eq. (4.46), holds. For $t \geqq T$, the Region II solution, Eq. (4.51), holds.

4-1.3.2. Use of the Method of Characteristics for Complex Transients

Application of the previous approach requires a knowledge of the coolant flow as a function of time. This presupposes transient conditions such that the flow through the system is unaffected by the fluid properties. In many transient situations, particularly in a LOCA, we cannot assume this decoupling and must simultaneously consider the conservation of energy, momentum, and mass.

If we assume one-dimensional, homogeneous flow, we can write the conservation-of-mass equations as

$$\rho \frac{\partial V}{\partial z} + \frac{\partial \rho}{\partial t} + V \frac{\partial \rho}{\partial z} = 0 \ , \tag{4.53}$$

and the conservation-of-momentum equation as

$$\frac{\partial V}{\partial t} + V \frac{\partial V}{\partial z} + \frac{g_c}{\rho} \frac{\partial P}{\partial z} + g_c F = 0 \ , \tag{4.54}$$

where

V = fluid velocity, ft/s

F = frictional head loss per foot of length, lb force/lb mass ft

P = pressure, lb/ft^2,

and the other symbols are as previously defined. The gravitational effect (elevation head) has been omitted from Eq. (4.54).

The $\partial P/\partial z$ can be replaced by recalling that the velocity of sound, a_1, can be obtained from

$$a_1^2 = g_c \partial P / \partial \rho \ . \tag{4.55}$$

For the momentum balance, we then have

$$\frac{\partial V}{\partial t} + V \frac{\partial V}{\partial z} + \frac{a_1^2}{\rho} \frac{\partial \rho}{\partial z} = 0 \ . \tag{4.56}$$

In adiabatic single-phase flow, we would have $a_1^2 = \rho^{\gamma'-1}$, where $\gamma' = c_P/c_V$. In actuality, two-phase flow is encountered under nonadiabatic conditions. At any given pressure, a_1 is a function of ρ; therefore, over a narrow pressure and enthalpy range, we can make the approximation

$$a_1^2 = \rho^K \ , \tag{4.57}$$

where K is constant. We then can transform the conservation-of-mass and conservation-of-momentum equations to

$$a_1 \frac{\partial V}{\partial z} + \frac{2}{K} \left(\frac{\partial a_1}{\partial t} + V \frac{\partial a_1}{\partial z} \right) = 0 \tag{4.58}$$

$$\frac{\partial V}{\partial t} + V \frac{\partial V}{\partial z} + \frac{2a_1}{K} \left(\frac{\partial a_1}{\partial z} \right) + g_c F = 0 \ . \tag{4.59}$$

Our energy equation is now more complex than previously, since ρ can no longer be assumed constant, and we obtain

$$\frac{q''p'}{A} = \frac{\partial}{\partial z} \left[G \left(H + \frac{v^2}{2g} \right) \right] + \frac{\partial}{\partial t} \left[\rho \left(U + \frac{v^2}{2g} \right) \right] \ . \tag{4.60}$$

If we replace U in terms of H, ρ, and P and use the conservation-of-mass equation, Eq. (4.60) can be transformed to

$$\frac{q''p'}{A\rho} - \frac{1}{J\rho}\left(\frac{\partial P}{\partial t}\right) = \frac{\partial H}{\partial t} + V\frac{\partial H}{\partial z} + \left[\frac{V}{2g_c}\left(\frac{\partial V^2}{\partial z}\right) + \frac{1}{2g_c}\left(\frac{\partial V^2}{\partial t}\right)\right] ,$$ (4.61)

where J equals the mechanical equivalent of heat, Btu/(ft lb). If the work due to friction is considered negligible in relation to the other terms in Eq. (4.61), the sum within the brackets at the right can be shown to be zero. Therefore,

$$\frac{q''p'}{A\rho} - \frac{1}{J\rho}\left(\frac{\partial P}{\partial t}\right) \approx \frac{\partial H}{\partial t} + V\frac{\partial H}{\partial z} .$$ (4.62)

We observe that the three conservation equations are quasilinear, partial-differential equations similar to those for which we found the method of characteristics applicable. Indeed, Courant and Friedrichs[22] have shown how the notion of characteristic directions can be extended to n quasilinear, partial-differential equations in two independent variables. Although an analytical solution is not possible under general conditions, Lister[23] showed how the method of characteristics can be used to obtain numerical solutions for flow transients, and Fabic[24] used the method for analysis of LOCAs.

Let us follow Lister[23] and consider Eq. (4.58), designated L_1, and Eq. (4.59), designated L_2, assuming a value for K is known. We form the linear combination L, where

$$L = \lambda_1 L_1 + \lambda_2 L_2 .$$ (4.63)

Therefore,

$$L = \frac{\partial V}{\partial z}(a_1\lambda_1 + V\lambda_2) + \frac{\partial V}{\partial t}(\lambda_2) + \frac{\partial a_1}{\partial z}(V\lambda_1 + a_1\lambda_2)(2/K) + \frac{\partial a_1}{\partial t}\left(\frac{2}{K}\lambda_1\right) + g_c F\lambda_2 = 0 .$$ (4.64).

We now choose values for λ_1 and λ_2 such that

$$\frac{dt}{dz} = \frac{\lambda_2}{a_1\lambda_1 + V\lambda_2} = \frac{\lambda_1}{V\lambda_1 + a_1\lambda_2} .$$ (4.65)

Next, we multiply both sides of Eq. (4.64) by dt

$$L dt = (\lambda_2)\left(\frac{\partial V}{\partial z}dz + \frac{\partial V}{\partial t}dt\right) + \left(\frac{2}{K}\lambda_1\right)\left(\frac{\partial a_1}{\partial z}dz + \frac{\partial a_1}{\partial t}dt\right) + g_c F\lambda_1 dt .$$ (4.66)

Since $V = V(Z,t)$ and $a_1 = a_1(z,t)$,

$$dV = \frac{\partial V}{\partial z}dz + \frac{\partial V}{\partial t}dt , \quad da_1 = \frac{\partial a_1}{\partial z}dz + \frac{\partial a_1}{\partial t}dt ,$$ (4.67)

therefore,

$$L dt = \lambda_2 dV + [(2/K)\lambda_1]da_1 + g_c F\lambda_1 dt .$$ (4.68)

In the expression for L in Eq. (4.64), the derivatives of V and a have been combined to be in the same direction. This direction (dz/dt) is called the "characteristic direction" or ζ. Equation (4.66) can be solved for the ratio λ_1/λ_2

$$-\frac{\lambda_1}{\lambda_2} = \frac{dz - Vdt}{-a_1 dt} = \frac{-a_1 dt}{dz - Vdt} \quad . \tag{4.69}$$

Hence,

$$dz^2 - 2Vdzdt + (V^2 - a_1^2)dt^2 = 0 \quad . \tag{4.70}$$

There are two real solutions for dz/dt,

$$\frac{dz}{dt} = \zeta_+ = (V + a_1) \; , \quad \frac{dz}{dt} = \zeta_- = (V - a_1) \quad . \tag{4.71}$$

Equation (4.71) defines two families of characteristic curves (designated C_+ and C_-) in the z,t plane. Since ζ_+ and ζ_- are functions only of z and t, we can substitute these into Eq. (4.69) and combine the results with Eq. (4.64) to obtain

$$dV + (2/K)da_1 - g_c Fdt = 0 \; , \tag{4.72}$$

and

$$-dV + (2/K)da_1 - g_c Fdt = 0 \tag{4.73}$$

as the two characteristic equations.

The energy equation must also be written as a total differential equation. We rewrite Eq. (4.62) to eliminate the pressure term, and obtain

$$\frac{q''p'}{A\rho} - \frac{a_1^2}{J\rho g_c}\left(\frac{\partial \rho}{\partial t}\right) = \frac{\partial H}{\partial t} + V\frac{\partial H}{\partial z} \quad . \tag{4.74}$$

If we were to temporarily assume that $(\partial \rho/\partial t)$ is a constant, Eq. (4.74) would be in the form of Eq. (4.36), and we then make use of Eq. (4.38) to obtain

$$dt = dz/V \tag{4.75}$$

or

$$V = dz/dt \quad . \tag{4.76}$$

Substituting for V in Eq. (4.74) yields

$$\left[\frac{q''p'}{A\rho} - \frac{-a_1^2}{J\rho g_c}\left(\frac{\partial \rho}{\partial t}\right)\right] = \frac{\partial H}{\partial t} + \left(\frac{dz}{dt}\right)\frac{\partial H}{\partial z} \quad . \tag{4.77}$$

On multiplication by dt, we find the right side to be the total differential, dH

$$\left[\frac{q''p'}{A\rho} - \frac{a_1^2}{J\rho g_c}\left(\frac{\partial \rho}{\partial t}\right)\right]dt = dH \quad . \tag{4.78}$$

This is considered a third characteristic equation whose characteristic direction is equal to V.

Characteristic equations can also be written when the homogeneous flow assumption is removed, providing some simplifications are made. Tentner and Weisman[25] indicate one approach that can be used. The conservation equations are written in terms of liquid velocity, V. Gas velocity, V_g, is taken as SV where S is the slip ratio. For simplicity of notation, they write

$$\frac{d\rho_g}{dp} = \dot{a}^2 \; ; \quad \frac{dH_g}{dp} = n^2 ; \quad \frac{dH_l}{dp} = m^2 \; , \tag{4.79}$$

where

ρ_g = gas density

P = pressure

H_g, H_l = gas and liquid enthalpies, respectively.

They then make the substitutions

$$\frac{\partial \rho_g}{\partial t} = \dot{a}^2 \frac{\partial P}{\partial t} \; ; \quad \frac{\partial \rho_g}{\partial z} = \dot{a}^2 \frac{\partial P}{\partial z} \; ; \quad \frac{\partial H_g}{\partial t} = n^2 \frac{\partial P}{\partial t}$$

$$\frac{\partial H_g}{\partial z} = n^2 \frac{\partial P}{\partial z} \; ; \quad \frac{\partial H_l}{\partial t} = m^2 \frac{\partial P}{\partial t}; \quad \frac{\partial H_l}{\partial z} = m^2 \frac{\partial P}{\partial z} \; , \tag{4.80}$$

where z is axial distance and t is time.

With the further assumption that $\frac{\partial \rho_l}{\partial t}, \frac{\partial \rho_l}{\partial z}, \frac{\partial S}{\partial t}, \frac{\partial S}{\partial z}$ are sufficiently small to be ignored, on rearrangement the conservation equations become

$$(\alpha \dot{a}^2) \frac{\partial P}{\partial t} + (\alpha S \dot{a}^2) \frac{\partial P}{\partial z} \, | \, 0 \frac{\partial V}{\partial t} + [\alpha \rho_g S + (1 - \alpha)\rho_l] \frac{\partial V}{\partial z}$$

$$+ (\rho_g - \rho_l) \frac{\partial \alpha}{\partial t} + (\rho_g SV - \rho_l V) \frac{\partial \alpha}{\partial z} = 0 \; , \tag{4.81}$$

$$(\alpha S V \dot{a}^2) \frac{\partial P}{\partial t} + (1 + \alpha S^2 V^2 \dot{a}^2) \frac{\partial P}{\partial z} + [\alpha \rho_g S + (1 - \alpha)\rho_l] \frac{\partial V}{\partial t}$$

$$+ [2\alpha \rho_g S^2 V + 2(1 - \alpha)\rho_l V] \frac{\partial V}{\partial z} + (\rho_g SV - \rho_l V) \frac{\partial \alpha}{\partial t} + (\rho_g S^2 V^2 - \rho_l V^2) \frac{\partial \alpha}{\partial z} = -\frac{\tau'}{A} \; , \tag{4.82}$$

$$[\alpha \dot{a} E_g - 1 + n^2 \alpha \rho_g + m^2(1 - \alpha)\rho_l] \frac{\partial P}{\partial t} + [\alpha S E g \dot{a}^2 + n^2 \alpha \rho_g VS + m^2(1 - \alpha)\rho_l V] \frac{\partial P}{\partial z}$$

$$+ [\alpha \rho_g S^2 V + (1 - \alpha)\rho_l V] \frac{\partial V}{\partial t} + [\alpha \rho_g S E_g + \alpha \rho_g V^2 S^3 + (1 - \alpha)\rho_l E_l + (1 - \alpha)\rho_l V^2]$$

$$\times \frac{\partial V}{\partial z} + (\rho_g E_g - \rho_l E_l) \frac{\partial \alpha}{\partial t} + (\rho_g V S E_g - \rho_l V E_l) \frac{\partial \alpha}{\partial z} = \frac{q'}{A} \; , \tag{4.83}$$

where

α = void fraction

ρ_l = liquid density

A = cross-sectional area

$$E_g = H_g + \frac{S^2 V^2}{2}$$

$$E_l = H_l + \frac{V^2}{2}$$

τ' = wall shear stress.

Each of the conservation equations is now in the general form

$$A_{i,1} \frac{\partial P}{\partial t} + A_{i,2} \frac{\partial P}{\partial z} + A_{i,3} \frac{\partial V}{\partial t} + A_{i,4} \frac{\partial V}{\partial z} + A_{i,5} \frac{\partial \alpha}{\partial t} + A_{i,6} \frac{\partial \alpha}{\partial z} + A_{i,7} = 0$$

$$i = 1, 2, 3 \quad , \tag{4.84}$$

which is suitable for solution by the method of characteristics.

Tentner and Weisman[25] find that real characteristics will be obtained providing we recognize that at high fluid velocities the slip ratio approaches unity. They use Hughmark's[26] slip ratio for $G < 1.8 \times 10^6$ lb/h ft². Since for G above $\sim 2 \times 10^6$ lb/h ft² the flow is believed homogeneous, S is taken as unity for $G > 2.2 \times 10^6$. For $1.8 \times 10^6 < G < 2.2 \times 10^6$, S is taken as a weighted average of the value obtained from the Hughmark correlation and 1.0. Failure to use slip ratios near unity at very high velocities leads to imaginary characteristics at low pressures. This would appear to be another confirmation of the conclusion that a two-phase mixture behaves as an essentially homogeneous fluid at very high mass velocities.

An alternative to slip ratio S for representing the relative motion of the vapor is the use of vapor drift velocity, V_{gm}, where

$$V_{gm} = V_g - V_m = (Q_g/\alpha A) - (Q_g + Q_l)/A \quad , \tag{4.85}$$

where

A = total flow area

V_g, V_m = gas and mixture velocities, respectively

Q_g, Q_l = gas and liquid volumetric flow rates

α = void fraction.

Kroeger[27] reports that the THOR code has taken this approach. Early versions successfully used the method of characteristics for solving the resulting conservation equations (later versions used a finite difference formulation). The THOR code also includes the effect of thermodynamic nonequilibrium between phases.

4-1.3.3. Numerical Solution Procedure for Characteristic Equations

Lister[23] indicates that numerical solutions can be obtained by using a grid of characteristics or by the method of specified time intervals. The latter method appears more easily adaptable to the complex geometries encountered in any real problem. We assume that the homogeneous flow assumption applies, that the system is divided into a series of axial segments of length ΔZ, and that we know the values of V, H, ρ, and a_1 at the grid locations at time t (points A, B, and C of Fig. 4-6). We want to find the values of the problem parameters at a given location and time, $t + \Delta t$ (point P of Fig. 4-6). Curves ζ_+ and ζ_- are two of the characteristic curves through P. Then, if R and S are the intersections of the characteristics with the horizontal line through time t, from Eq. (4.71) we have

$$(Z_p - Z_R) = \zeta_+(\Delta t)$$

$$(Z_p - Z_S) = \zeta_-(\Delta t) \ . \tag{4.86}$$

We need the values of V, a_1, (Z_R, t), and (Z_S, t), which can be estimated by linear interpolation from the values at A, C, and B. We then use a finite difference form of Eqs. (4.72) and (4.73) to obtain V and a_1 at point P:

$$(V_P - V_R) + (2/K)(a_{1P} - a_{1R}) - g_c F(t_p - t_R) = 0 \tag{4.87}$$

$$(V_P - V_S) + (2/K)(a_{1P} - a_{1R}) - g_c F(t_p - t_S) = 0 \ . \tag{4.88}$$

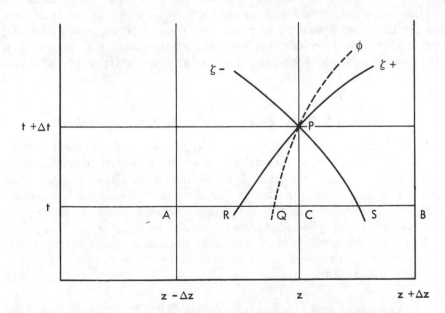

Fig. 4-6.   Characteristic curves with grid of specified time intervals.

The value of a_{1p} is then used to obtain an updated value for ρ from Eq. (4.57), $\Delta\rho/\Delta t$, and $\Delta P/\Delta t$.

Our energy equation, Eq. (4.78), is the third characteristic ϕ through point P. Position Z_q at which this curve intersects the line through t is obtained from the finite difference form of Eq. (4.76).

$$Z_p - Z_q = V(\Delta t) \ . \tag{4.89}$$

The enthalpy at point Q is obtained by linear interpolation and used to obtain H_p

$$H_p = \left[\frac{q''p'}{A\rho} - \frac{a_1^2}{J\rho g_c}\left(\frac{\Delta\rho}{\Delta t}\right)\right]\Delta t + H_Q \ . \tag{4.90}$$

New values of ρ, P, and H can then be used to update K and, if necessary, the calculation can be repeated using an improved average value of K. If the change has been small, the calculation proceeds to the next grid point to be evaluated.

The characteristic curves have a direct physical interpretation. The ζ curves can be thought of as compression and rarefaction waves traveling through the system at near sonic velocities; the ϕ curve represents the transport of thermal energy through the system, which occurs at a rate determined by fluid velocity. Once we have divided the system into a series of increments, we are no longer completely free in our choice of time step Δt, and it must be such that

$$\Delta t \leqq \Delta Z/(a_{\max}) \ , \tag{4.91}$$

where $a_{\max}$ represents the maximum value the acoustic velocity has anywhere in the system. Without this restriction, we could not obtain the properties on the characteristic curve at time t by interpolation between the properties at adjacent nodes. In this interpolation, we assumed that the characteristic curves are linear, while in reality they are not. For improved accuracy, quadratic interpolation procedures can be used[23] for determining conditions at points R, Q, and S (see Fig. 4-6).

4-1.3.4 Numerical Solution Procedures Via Difference Methods

Although use of a characteristic mesh provides a satisfactory means for solving the partial differential equations resulting from fluid transients, its obvious complexity has led to other approaches. These alternative integration procedures are generally based on differences taken in the direction of the original independent variables. These methods have the advantage that solution values can be found at evenly spaced points in the solution plane.

Basic conservation equations all express the time rate of change of one variable in terms of spatial changes in the others. Generally, those can be expressed as first-order equations of the form

$$\partial u_i/\partial t = \sum_k [A_k(\partial u_k/\partial x) + B_k u_k] \ , \tag{4.92}$$

where u_k are variables other than time, t, and space, x, and A_k and B_k depend on x, t, and u_k. When expressed in finite difference form, the time rate of change can be approximated by the spatial change observed in the previous time step. That is, for a particular variable we write

$$\frac{u_i^{j,n+1} - u_i^{j,n}}{\Delta t} = \sum_k \left[A_k \left(\frac{u_k^{j+1,n} - u_k^{j-1,n}}{\Delta x} \right) + B_k(u_k^{j+1,n} - u_k^{j-1,n}) \right], \qquad (4.93)$$

where notation $u_k^{j,n}$ indicates the value of variable u_k at spatial location j and time step n. With this approach, the values of independent variables at time period $(n + 1)$ can be directly determined from the values at time n. This approximation procedure requires that the time step be restricted so that

$$\Delta t \leqslant \Delta x / (a_1 + |V|), \qquad (4.94)$$

where V is fluid velocity and a_1 is sonic velocity. The general conditions for stability and convergence of finite difference schemes of this nature are discussed by Fox.[28]

The time step restriction imposed on the previously described explicit method of Eq. (4.94) can lead to excessive computing times. This difficulty can be circumvented by going to a fully implicit method. Here spatial changes are written in terms of variable values at the current time step. That is, we write equations of the form

$$\frac{u_i^{j,m+1} - u_i^{j,n}}{\Delta t} = \sum_k \left[A_k \left(\frac{u_k^{j+1,n+1} - u_k^{j-1,n+1}}{\Delta x} \right) + B_k(u_k^{j+1,n+1} - u_k^{j-1,n+1}) \right]. \qquad (4.95)$$

This approach requires the solution of a set of simultaneous equations covering all the variables at all spatial locations. However, the time step limitation of Eq. (4.94) is removed [accuracy of the solution requires $\Delta t \leqslant x/(|V|)$]. Procedures have been developed for minimizing the effort involved in solving these simultaneous equations.[29]

It was observed that the method of characteristics was based on the concept of using a series of control volumes for representing the system and writing the conservation-of-mass, -energy, and -momentum equations for each control volume. Although difference procedures can also be based on the use of control volumes, they have more usually been based on the "node and branch" concept.

In electrical and mechanical systems, it is usual to represent a distributed parameter system by a series of nodes connected to each other through mechanical or electrical resistances. When applied to fluid dynamics, it is assumed that we have a series of spacial elements (nodes) with a capability for energy and mass storage only. The nodes are connected by branches that contain flow resistance and inertance. Thus, the conservation-of-momentum equation is written for each branch, while the conservation-of-energy and -continuity equations are written for each node. Nahavandi[30] and Murphy et al.[31] described computational procedures of this type. In the FLASH codes,[31] the conservation equations are written in the form

(a) conservation of nodal energy

$$\frac{U_i(t + \Delta t) - U_i(t)}{\Delta t} = \sum_j w_{ij} H_{ij} + Q_i \ , \tag{4.96}$$

(b) conservation of nodal mass

$$\frac{M_i(t + \Delta t) - M_i(t)}{\Delta t} = \sum_j w_{ij} \ , \tag{4.97}$$

(c) conservation of flowpath (branch) momentum

$$(1/A)(\Delta w_{ij}/\Delta t) = (g_c/L_j)\left(P_{i+1} - P_i - \frac{C_j |w_{ij}| w_{ij}}{2 g_c A^2 \rho} + \rho Z_j\right) , \tag{4.98}$$

where

A = flowpath area

H_{ij} = enthalpy associated with flow w_{ij}, Btu/lb

C_j = pressure loss coefficient for branch j

Q_i = total heat generation rate in node i, Btu/s

M_i = total mass of node i

L_j = length of flowpath (branch) j

P_i = pressure at node i, lb$_f$/ft^2

U_i = total fluid internal energy in control volume i, Btu

w_{ij} = flow into node i through path j, lb/s

Z_j = difference in elevation head along branch j.

Note that the spatial differences are taken between adjacent nodes. Further, the flow in a given branch is taken as constant along that branch and the effects of spatial acceleration thus are not incorporated in the model.

The RELAP codes,[32] which are based on FLASH, use essentially the same basic equations. Early versions of FLASH (Ref. 31) used an explicit integration procedure, while later versions used the implicit approach.[33] When the implicit approach is used with a time step significantly larger than pressure wave transport time, the fine structure of the acoustic effects is suppressed. Such a solution is similar to one obtained when pressure wave propagation is assumed to be instantaneous.

4-1.3.5 Accelerated Solution Techniques

All the foregoing techniques require considerable computer time for most transients. A number of approaches have been taken to reduce the time needed for calculating system behavior in slowly varying transients where acoustic effects can be ignored.

Agee[34] proposed an approach where the finite difference equations, Eqs. (4.96), (4.97), and (4.98), are replaced by integral forms in which the parameters are assumed to vary along the length step. A modified set of momentum equations are solved implicitly; the results are then used to obtain explicit solutions of the other two conservation equations. Agee claims that this approach can allow meaningful results to be obtained with substantially longer time steps than are useful with an unmodified implicit solution procedure.

A more common approach to slowly varying system transients (e.g., pump coastdown) is to assume that flow in the system can be obtained from a knowledge of pump behavior (e.g., use of a pump coastdown curve) and then to solve only the conservation-of-energy and -mass equations for each control volume. Elimination of the momentum equation results in a substantial decrease in computing time. The DYNODE code is an example of this technique.

A somewhat related approach is used by the COBRA codes[9] for examining core behavior during a transient. Rowe[9] writes the basic conservation equations for an axial segment (length Δz) of subchannel i as

Continuity

$$A_i(\partial \rho_i/\partial t) + (\partial m_i/\partial z) = -w_{ij} \tag{4.99}$$

Energy

$$(1/V_i')(\partial H_i/\partial t) + (\partial H_i/\partial z)$$

$$= (q_i'/m_i) - (H_i - H_j)(w_{ij}'/m_i) - (T_i - T_j)(k_{ij}/m_i) + (H_i - H^*)(w_{ij}/m_i) \tag{4.100}$$

Axial Momentum

$$(1/A_i)(\partial m_i/\partial t) - 2V_i(\partial \rho_i/\partial t) + (\partial P_i/\partial z)$$

$$= -(m_i/A_i)^2 \left[\frac{f_i \phi^2}{2\rho_l D_e} + \frac{A_i \partial(1/\rho_i A_i)}{\partial z} \right] - \rho_i g \cos \theta$$

$$- (f_t/A_i)(V_i - V_j)w_{ij}' + (1/A_i)(2V_i - V^*)w_{ij} \tag{4.101}$$

Lateral Momentum

$$(\partial w_{ij}/\partial t) + \frac{\partial(V_{ij}^* w_{ij})}{\partial z} + (s/l)(C_{ij} w_{ij}) = (s/l)(P_i - P_j) \tag{4.102}$$

where

A_i = flow area of channel i

C_{ij} = loss coefficient for transverse flow

D_e = hydraulic diameter

f_i = friction factor based on all liquid flow

f_t = turbulent momentum factor

g = gravitational constant

H_i, H_j = enthalpy of fluid in subchannels i and j, respectively

H^* = enthalpy carried by cross flow

k_{ij} = thermal conductivity

l = effective length over which lateral momentum transfer takes place (see Chap. 3)

m_i = flow rate in channel i, mass/time

P_i, P_j = pressure in channels i and j, respectively

q_i' = heat addition per unit length in channel i

s = gap between adjacent rods

t = time

T_i, T_j = temperature of fluids in channels i and j, respectively

V_i' = effective enthalpy transport velocity in channel i, where $V_i' = V_i$ for homogeneous flow

V_i, V_j = axial velocities in channels i and j, respectively

V^* = effective axial velocity carried by cross flow

w_{ij} = net cross flow between channels i and j

w_{ij}' = turbulent cross flow between channels i and j (accounts for mixing, no net transfer of mass)

ϕ^2 = two-phase friction multiplier

ρ_l, ρ = densities of liquid and mixture, respectively.

Note that Eq. (4.102) is based on the assumption that transverse momentum flux terms can be ignored.

Total flow to the core is assumed given. Since flow distribution between the channels is to be determined, axial and lateral momentum balances are required. However, omission of a $\partial \rho / \partial t$ term from the momentum balance means that sonic velocity propagation effects are ignored. Therefore, there is no requirement that very small time steps be used with an explicit solution procedure.

4-1.3.6 Advanced Computational Techniques

The computational techniques previously described have all made the simplifying assumptions that (a) flow can be considered one dimensional, and (b) a single set of conservation equations can describe average liquid and vapor behavior. A more rigorous approach recognizes that flow through some portions of the reactor system is multidimensional and behavior of the liquid and vapor is more properly described by writing separate conservation equations for each phase. Numerical techniques capable of dealing with such complex systems have been devised and used to examine two-phase behavior in limited regions.

Harlow and Amsden[35,36] devised the so-called Implicit, Multi-Field (IMF) Technique that is applicable when studying time-varying (initial-value) problems in several space dimensions with a continuous and dispersed phase. To avoid numerical instability, they use an implicit formulation of the finite difference equations which incorporates viscous dissipation effects.[37] The computer technique includes provisions for visual displays of the output as originally developed for the MAC code.[38]

Multidimensional, multifield calculations can be expected to be considerably more time consuming than the one-dimensional and single-field techniques now (1977) in general use. Therefore, these techniques are not likely to replace present techniques rapidly, but rather to provide an improved understanding of system behavior. This increased understanding would then be reflected in the development of more realistic approximations in the codes routinely used for system analysis.

4-2 FORCED CONVECTION HEAT TRANSFER IN SINGLE-PHASE FLOW

Since we established the means for determining fluid enthalpy along the reactor core, we are now able to determine existing heat-transfer regimes. Single-phase forced convection heat transfer is encountered in the inlet and low-power regions of a PWR core where fluid enthalpy is below that of the saturated liquid. It is also encountered on the tube side of the steam generators. Conditions for the onset of boiling are described in detail in Sec. 4-3.

4-2.1 Empirical Equations for Single-Phase Heat Transfer

While laminar flow ($Re < 2000$) will never be encountered during any expected operating situation, some calculations of hypothetical accident conditions indicate that this regime can be found. We can then follow Rohsenow and Choi's[39] modification of the theoretical equation and obtain the heat transfer coefficient from

$$hD_e/k = 4.0 \ . \tag{4.103}$$

Despite marked progress in understanding turbulence, the limitations and complexities of theoretical approaches have led to the use of empirical correlations in turbulent flow. For turbulent flow of water inside smooth conduits and annuli, the most extensively used correlation is that of Dittus and Boelter[40]:

$$\left(\frac{hD_e}{k}\right)_b = C \left(\frac{D_e G}{\mu}\right)_b^{0.8} \left(\frac{c_p \mu}{k}\right)_b^n \ , \tag{4.104}$$

where

h = heat transfer coefficient, Btu/(h ft^2 °F)

D_e = equivalent diameter, ft

k = thermal conductivity of fluid, Btu/(h ft °F)

c_p = specific heat of fluid, Btu/lb

G = mass velocity, lb/(h ft^2)

μ = fluid viscosity, lb/(h ft)

C = 0.023

n = 0.4

$\left(\dfrac{D_e G}{\mu} \right) > 10\,000$ and $L/D_e > 60$

b = bulk conditions.

The Dittus-Boelter correlation is also widely used for water flow parallel to tube banks. Available experimental data indicate that the coefficient of Eq. (4.104) varies with the pitch-to-diameter ratio. The h values obtained from the Dittus-Boelter correlation are slightly conservative for the pitch-to-diameter ratios of interest to reactor designers. Weisman[41] correlated the available data for triangular-pitch lattices using Eq. (4.104) with

$$C = 0.026(s/D) - 0.006$$

and square-pitch lattice data with

$$C = 0.042(s/D) - 0.024 \ , \tag{4.105}$$

where

s = tube pitch

D = tube diameter.

The results for both types of lattices can be expressed as

$$C = 0.0333\,E + 0.0127 \ , \tag{4.106}$$

where E is the fraction of the total cross-sectional area in an infinite array taken up by the fluid. Variations in the heat coefficients around the periphery of a rod have been found negligible for s/D ratios >1.2 (Ref. 42).

While all present PWR cores are designed with flow parallel to the fuel elements, some consideration has been given to the use of cross flow. Cross flow, which provides higher heat transfer coefficients than attainable at the same mass flow rate in parallel flow, is now encountered on the shell side of a once-through steam generator. The heat-transfer data for flow of liquids normal to banks of